HERETIC

HERETIC

ONE SCIENTIST'S
JOURNEY FROM
DARWIN TO DESIGN

MATTI LEISOLA

JONATHAN WITT

SEATTLE   DISCOVERY INSTITUTE PRESS   2018

Description

What happens when an up-and-coming European bioscientist flips from Darwin disciple to Darwin defector? Sparks fly. Just ask biotechnologist Matti Leisola.

It all started when a student loaned the Finnish scientist a book criticizing evolutionary theory. Leisola reacted angrily, and set out to defend evolution, but found his efforts raised more questions than they answered. He soon morphed into a full-on Darwin skeptic, even as he was on his way to becoming a leading bio-engineer.

Heretic is the story of Leisola's adventures making waves—and many friends and enemies—at major research labs and universities across Europe. Tracing his investigative path, the book draws on Leisola's expertise in molecular biology to show how the evidence points more strongly than ever to the original biotechnologist—a designing intelligence whose skill and reach dwarf those of even our finest bioengineers, and leave blind evolution in the dust.

Library Cataloging Data

Heretic: One Scientist's Journey from Darwin to Design by Matti Leisola and Jonathan Witt

258 pages, 6 x 9 x 0.54 in. & 0.78 lb, 229 x 152 x 14 mm & 351 g

Library of Congress Control Number: 2018930034

SCI027000 SCIENCE/Life Sciences/Evolution

BIO015000 BIOGRAPHY & AUTOBIOGRAPHY/Science & Technology

SCI008000 SCIENCE/Life Sciences/Biology

ISBN-13: 978-1-936599-50-9 (paperback), 978-1-936599-49-3 (Kindle), 978-1-936599-51-6 (EPUB)

Publisher Information

Discovery Institute Press, 208 Columbia Street, Seattle, WA 98104

Internet: http://www. discoveryinstitutepress.org/

Published in the United States of America on acid-free paper.

First Edition: February 2018.

Endorsements

"Award-winning Finnish biotechnologist Matti Leisola has written a fascinating account of what happens when a scientist follows the evidence wherever it leads. Leisola's account of how he succeeded should inspire up-and-coming scientists who face the same challenge."

Biologist Jonathan Wells, PhD,
author of *Icons of Evolution* and *Zombie Science*

"Scientists, like all other intellectuals, have ideas about what constitutes and what does not constitute reality. However, they are often not aware—and sometimes not ready to admit—that such ideas represent the principles of their philosophy. Leisola and Witt's *Heretic* is a unique first-hand account of the life-long adventures of a scientist who dared to challenge philosophical principles of colleague scientists. In my opinion, the outcome shows that to many scientists their philosophy is dearer than their science."

Biochemist and inventor Branko Kozulić, PhD

"This book is an exciting story about how a scientist's relentless search for truth makes him a heretic in the eyes of a cultural community more concerned about prestige than principle."

Tapio Puolimatka, PhD and EdD, University of Jyväskylä, Finland

"This book is a personal, strong, and motivated plea for intelligent design (ID) and 'swims against the current' of Darwinian evolution, now generally accepted in scientific circles and society. I personally do not endorse ID, but I am a good friend of the author, whom I also highly respect as a scientist active in academia and in the biotech industry over so many years. *Heretic* inspires readers to think critically and to open up a civilized discussion on neo-Darwinism versus ID. It covers the science and philosophical parts adequately; it is accessible to a large readership; and statements are underpinned by relevant research and literature data. Its value lies in the author's lifelong engagement and personal crusade to stimulate the public debate among scientists as well as laymen over Darwinism (chance/random mutation and natural selection) versus ID, a vision that Leisola strongly advocates.

Dr. Erick J. Vandamme, Emeritus Professor of Bioscience
Engineering, Centre for Biotechnology and Synthetic
Biology, Ghent University, Ghent, Belgium

"Matti Leisola has written the exciting story of almost the entire spectrum of aberrant motives, absurd fears, and unreasonable reactions to intelligent design (ID) by evolutionary scientists, clergymen, and church institutions alike, notably during his career as a scientist over the last some forty years. I would add a word on the fears of so many critics that accepting ID also means accepting the dogmata of some 1700 years of church history. ID is thoroughly neutral concerning such topics. So, the reader is invited to carefully check the historical and, what is more, the enormous wealth of scientific data Matti Leisola has presented in the present book: Test them carefully with an open mind and form your own independent opinion!"

Dr. Wolf-Ekkehard Lönnig, geneticist, Cologne, Germany,
author of many books about intelligent design, his latest one
about dogs and macroevolution (in German, 2014)

CONTENTS

ACKNOWLEDGMENTS

HERETIC IS A JOINTLY AUTHORED WORK, BUT IT IS MATTI LEISO-la's story, so the first-person singular refers to Matti throughout the book.

Both authors want to thank John West and Discovery Institute's Center for Science and Culture for taking on this project, and in particular the input from CSC Senior Fellow Jonathan Wells. Also, grateful thanks for helpful feedback from Ann Gauger, Pat Achey, and Amanda Witt, endnote help from Sarah Chaffee, and able manuscript production work by Rachel Adams and Mike Perry.

This is a substantially new book, but its foundation is Matti's earlier Finnish work, *Evoluutiouskon Ihmemaassa*. Matti wants to thank his friends Pekka Reinikainen, Tapio Puolimatka, Leena Laitinen, and Reino Kalmari, for their constructive comments. Lennart Saari filled in some gaps about the seminar described in Chapter 4. Kimmo Pälikkö helped with the pictures. Matti is also in debt to numerous colleagues for stimulating, constructive, and sometimes vigorous discussions. He especially wants to thank his research colleagues Dr. Ossi Pastinen and the protein chemist Dr. Ossi Turunen.

Artist Ray Braun granted permission for repurposing some of his diagrams from *Darwin's Doubt* by Stephen Meyer. *Darwin's Doubt* is mentioned at several points in this book, but we have summarized some of its ideas in chapters 2, 5, 6, and 12 without providing a citation at every point.

There are others who provided encouragement at crucial moments early in Matti's awakening. He looks forward to thanking them in person one day, in the words of C. S. Lewis, "further up and further in."

Introduction

As a scientist in my native Finland and later in Switzerland, I have been privileged to take part in some significant scientific breakthroughs, to lead some groundbreaking research projects in biochemistry and biotechnology, and to work alongside some world-famous scientists from Europe, Japan, and the United States. While we did not always see eye to eye, we shared a love for scientific testing and discovery. But there was another side of me, as there was and is another side to contemporary scientific culture.

As a young student, I used to laugh at those who, as I thought, *placed God in the gaps of our scientific knowledge*. This God-of-the-gaps criticism is often leveled against Christians and other religious believers, against all those who insist there is clear evidence of design in nature. To my way of thinking, such people lacked the patience and level-headedness that I possessed. It was so clear to me: Instead of plugging away to discover the natural mechanism for this or that mystery about the natural world, these pro-design people threw up their hands and used the God-did-it explanation as a cover for ignorance. This criticism of intelligent design proponents struck me as reasonable, so I didn't listen to their arguments.

But eventually I came to realize that this criticism cuts both ways, since a functional atheist also can reach for pat explanations in the face of mystery. It's just that for him, the pat explanation will never be God. That is, you do not need God in your explanatory toolkit in order to short-circuit careful scientific investigation and reasoning. I realized that I myself had been all too willing to stuff vague materialistic explanations into the gaps of our scientific knowledge.

I also recognized something else that the god-of-the-gaps criticism obscures: The more we learn about the natural world, the more fresh mysteries open up before us. David Berlinski, who has taught at Stanford, Rutgers, the City University of New York, and the Université de Paris, put it this way in *The Devil's Delusion*:

> Western science has proceeded by filling gaps, but in filling them, it has created gaps all over again. The process is inexhaustible. Einstein created the special theory of relativity to accommodate certain anomalies in the interpretation of Clerk Maxwell's theory of the electromagnetic field. Special relativity led directly to general relativity. But general relativity is inconsistent with quantum mechanics, the largest visions of the physical world alien to one another. Understanding has improved, but within the physical sciences, anomalies have grown great, and what is more, anomalies have grown great *because* understanding has improved.[1]

The god-of-the-gaps criticism ignores this well-established pattern. Materialists think that because we continue to make discoveries about the natural world, the pool of known mysteries must be shrinking toward zero. Instead, whole landscapes of new mystery present themselves to science precisely when some major new discovery is achieved, like the explorer reaching the crest of a mountain and finding a new realm before him.

Also, their argument for entertaining only material explanations in the sciences just assumes that everything we find in nature has a purely material cause. But what if that assumption is wrong? What if there are features of the natural world—the laws and constants of nature itself, for instance—that really are the work of a creative intelligence?

Scientists are supposed to investigate mysteries with an open mind, not assume an explanation from the outset. I came to see that the best approach is to evaluate which explanation among the live options is more logical and fits the facts better.

I also realized that it isn't very scientific to simply trust the majority of scientific specialists on a subject. The majority view may be correct, but the history of science shows that often it is incorrect. Scientific progress requires some healthy skepticism. And that means resisting the cult of "Science Says" even when authoritative sources reassure us with talk of "scientific studies."

One for-instance: In 1964 *The New York Times* reported that hundreds of scientific studies showed that there was no conclusive evidence that smoking causes lung cancer:

> Studies made in the last ten years have found no laboratory evidence linking lung cancer or fatal heart disease with cigarette smoking, The Council for Tobacco Research asserted yesterday in its 1963–64 report.
>
> A 71–page booklet written by Dr. Clarence Cook Little said the council had evaluated 350 reports by scientists working with council grants and had found "little to support" the charge that cigarette smoke produces cancer.[2]

Fortunately, in this case there were studies and official scientific voices pushing in the other direction, and soon the Council for Tobacco Research's 1963–64 report had lost credibility. Sometimes, however, the most prominent voices, the ones deemed to speak most authoritatively for science, have lined up squarely behind a position later found to be false. And such instances are not restricted to long centuries ago, when science was in its infancy.

The illness known as pellagra reached epidemic proportions in the U.S. in the early twentieth century. The scientific consensus attributed it to an infectious agent or moldy corn. It turned out to be a vitamin deficiency.[3] Also, through much of the twentieth century, the conventional scientific wisdom was that the continents were fixed. When German geologist Alfred Wegener published *Die Entstehung der Kontinente und Ozeane* (*The Origin of Continents and Oceans*), arguing for the idea of continental drift, he was excoriated as a sloppy crank, a man so enamored of his pet theory that he was blind to the facts. Fully half a century after his book appeared, he was still viewed with suspicion. Today his idea of drifting continents is the standard one in geology.

A third instance, even more recent: Backed by the authority of the U.S. government, the scientific establishment deemed eggs bad for your heart and pushed this narrative for years. These scientific authorities insisted eggs were bad for you, and ended up with egg on their faces.[4]

These examples underscore what should be obvious: How could science progress if we could never question or abandon the majority scientific opinion? We would all still be geocentrists who thought the continents were fixed and that eggs were terrible for you.

Swimming against the current isn't easy, of course. My own voyage away from the naturalistic evolutionary faith was long and painstaking. In this book, I describe that journey. I also detail the evasions, hatred, suspicions, contempt, fears, power games, and persecution that face scientists who oppose the evolutionary paradigm and the naturalistic worldview behind it.

I speak from firsthand experience. Over and over again I have encountered materialist fanaticism from people who are not ready to give up their views in the face of contrary evidence. Actually, they usually are not even interested in considering the evidence.

More on this later. Here a single example will suffice. In 2012 the results of a project called ENCODE[5] were published in the journal *Nature*. The name of the project comes from the words *Encyclopedia of DNA Elements*. Since 1970 leading evolutionists had claimed that most of the human genome is garbage left over from the random mutations said to fuel the evolutionary process. (See Chapter 8.) But the ENCODE project showed that the great majority of our genome is transcribed into RNA, suggesting that it is functional. These results were inconvenient to neo-Darwinists, but rather than objectively assessing the new findings and returning to the drawing board, many Darwinists responded with knee-jerk, sarcastic dismissals of the ENCODE results. The tone of the following passage is indicative (emphasis added):

> This claim *flies in the face* of current estimates... *This absurd conclusion* was reached through various means... Here, we detail the many logical and methodological *transgressions* involved in assigning

functionality to almost every nucleotide in the human genome. The ENCODE results were predicted by one of its authors to necessitate the rewriting of textbooks. We agree, many textbooks dealing with *marketing, mass media hype, and public relations may well have to be rewritten.*[6]

Such reactions are telling. They suggest that many neo-Darwinists are ready to dismiss out of hand any observations that do not fit their theory.

In science we should follow the evidence, not cling to pet theories. Nobel Laureate Richard Feynman well described the scientific ideal. "If it disagrees with experiment it is wrong," he commented. "In that simple statement is the key to science. It does not make any difference how beautiful your guess is. It does not make any difference how smart you are, who made the guess, or what his name is—if it disagrees with experiment it is wrong. That is all there is to it."[7]

Wise words, but easier said than done. This book is the story of where the evidence led me after I decided to follow it into the wilderness of heresy, and of the battles I was drawn into along the way.

The journey, understand, hasn't been all blood and bruises. I started out as a scientist in my native Finland, then spent several years as a scientist in Zürich, with many wonderful experiences in both places. I then returned to Finland in 1988 and worked as a research director in the biotech industry and then for fifteen years as a professor of bioprocess engineering. During this period I lectured on biotechnology, but also gave talks in several Finnish universities about chemical and biological evolution, with titles such as "Evolution: A Modern Creation Myth," "What Differentiates Men from Stone?" "From Stone to Man," and "The Riddle of the Origin of Life." The lecture halls were often full to bursting.

During these years I have spoken throughout Europe, North America and Japan, in polytechnic schools, scientific meetings, and Rotary clubs, at private research companies and public universities, often to highly engaged and animated audiences. But the most interesting and rewarding visits I have had were in high schools, and this despite the fact that the headmasters are not always excited about my visits, since I challenge some of the material in their assigned textbooks. While a research director at Cultor, I visited a nearby high school many times from 1991 to 1996. The students were more than interested listeners and discussion partners. One teacher expressed astonishment that when the lunch bell rang during my talk, the students did not bolt for the cafeteria, but instead continued listening and asking questions. One of the students later became one of my students and completed a doctoral thesis under my supervision.

This should not have surprised the teachers. Teaching biology (or any field of science) as settled dogma, and a dogma moreover that points to a universe drained of meaning and purpose—that is an approach hardly calculated to fascinate and draw young people into the sciences. But imagine teaching biology and other disciplines, like physics and astronomy, so that students are encouraged to think critically about scientific theories. Imagine the students being exposed not only to the evidence for a reigning theory but also to evidence contradicting the theory. And imagine the students not being force-fed one worldview masquerading as science but being freed to consider which worldview the evidence best supports. That's an approach all but guaranteed to energize and excite.

1. Suspicions Awoken

In 1972 I was sitting in the major lecture hall of the University of Helsinki as a young student of biochemistry. The American theologian and philosopher Francis Schaeffer had come to Helsinki to speak, and in the course of his lectures I realized how naive my concept of truth was. I went out and bought several of Schaeffer's books and started my reading in philosophy, which previously I had considered of little value.

To understand my thinking up to that time, you have to understand something about the culture I was raised in. True, from when I began primary school in 1954 to when I finished high school in 1966, Finland was almost completely a Lutheran country, and about 95% of the population belonged to the church. But if that is all you know, you will get the wrong idea about my childhood education. In school both the Bible and a naturalistic understanding of life's history were taught, but they were not presented on an equal footing. The Christian faith had a place in the schools, but just as Schaeffer had said in his books and lectures, it was presented as belonging to the area of nonreason. And what was thought to lie on the side of reason? It was assumed that rational thinking shows that man is only a machine produced by random processes.

Schaeffer called this chasm between reason and faith *the line of despair,* which he said became a fixture in Europe by around 1890. As he described this during the talk that I attended in a crammed lecture hall in March 1972, I immediately recognized the reality of it in my own culture. One could agree with the line of despair. One could disagree with it. But there was no denying that it dominated our way of thinking.

What also struck me at the time was how emotional those of us who accepted the line of despair could get when the line was challenged. After all, we were on the side of reason, weren't we? But sitting there listening to Schaeffer, I was no longer so sure. I thought back to a time three years before. I still remember it vividly. A student gave me a book written by an Indian scientist who was critical of evolution. The term "evolution," by the way, can refer to many things. For simplicity and unless otherwise stated, "evolution" and "the theory of evolution" in this book will refer broadly to the idea of the common descent of all living organisms from one or a very few common ancestors, incrementally diversifying through a blind, purely material process. This was the idea I had long accepted as fact, but this book I had been lent argued that it was far from fact— wasn't even well supported by the evidence. I reacted angrily. Who is this unknown Indian man anyway? Overwhelming evidence is against him, I reminded myself, and then I marshaled my then meager knowledge of biology to show that student just how wrong he was.

My effort, however, raised as many questions in my mind as it answered. And three years later, there in the lecture hall listening to Schaeffer, my emotional reaction to those who had challenged my faith in evolution struck me as suspicious. If I was the confident and rational one, why was I so touchy? Why did I react with anger when my understanding of the world was questioned? I came to see that emotional reactions are the rule rather than the exception when one's basic philosophical commitments are challenged. I started to wonder why I had cared so little about clear results and straightforward theoretical calculations that posed a challenge to my worldview. Why had I not even been interested in them?

I soon saw there was a key philosophical question bigger than science, and yet one that many insisted must be answered only in one way when thinking scientifically. In the *Philebus* of Plato (429–347 B.C.E.) Socrates considered this all-important question and laid out the two primary possibilities: "Whether we are to affirm that all existing things, and this fair scene which we call the Universe, are governed by the influence of the irrational, the random, and the mere chance; or, on the contrary, as our predecessors affirmed, are kept in their course by the control of mind and a certain wonderful regulating intelligence."[1]

The scientific establishment of our day does not, for the most part, explicitly argue for the former over the latter. Instead they simply insist that we must assume the former anytime we are doing science, must entertain only those explanations consistent with atheism, regardless of what we believe in our private lives. The name for this dogma is methodological materialism, and I came to realize how irrational this view of scientific rationality was.

Understand, most scientists who go along with methodological materialism put about as much thought into it as they do breathing. I was that way. And in hundreds of discussions over the years I have witnessed a blindness to basic philosophical commitments in many kinds of people from at least thirty different nationalities. Even among scientists few are aware of their basic presuppositions. Most of them consider science a neutral search for truth.

I realized this early in my journey away from Darwin. At one point I invited some of my science professors to discuss these issues with me. This was while I was at the Helsinki University of Technology. We met in my home. Three professors were present, and one started things off by ridiculing the Bible that was sitting on my table.

FIGURE 1.1—I started my studies in 1966 at the Helsinki University of Technology's Department of Chemistry and finished my career in the same building forty-six years later after spending seven years at Swiss Federal Institute of Technology (ETH) and nine years as a research director in a Finnish biotech company.

At one point in our conversation I asked them what the basis of their evolutionary thinking was, and very quickly I realized that they had no real answers. They took it all on faith, having swallowed the modern concept of truth without any further consideration. They had precious little evidence. What they had were stories.

The Origin of Life: Just So

ONE PILLAR of the materialist faith is an imaginative origins story about how the first life arose on planet Earth. Here is one telling of it:

> The early seas obviously contained large amounts of various organic compounds which came together and formed giant molecules. Some stable structures among those became slowly more common. Step by step through addition of new elements these molecules got new properties. They learned to harvest chemical energy from reactions for their own construction work. They could grow, divide and renew to similar structures. They started to get properties typical for living beings.[2]

Tens of thousands of Finnish high school students have studied this text from a 1974 biology textbook. It's a model example of a paradigm-controlled imaginary story with no basis in the known natural laws and principles of chemical reactions. Did the text cause a scandal? Did all of those hard-boiled mainstream scientists who demand hard evidence climb to the barricades in protest? Did the textbook get a hooey-prize from the community of evolutionary scientists? Nothing like that happened because the text fits the naturalistic view of life's origin.

Such imaginative storytelling about the origin of life, also referred to as *chemical evolution*, has a long history. A decade after *The Origin of Species* appeared, German evolutionist Ernst Haeckel made drawings depicting the spontaneous origin of life, including what he speculated was the reproductive cycle (See Figure 1.2) of imaginary single-celled organisms he called *Monera*.[3] What he depicted there had not been discovered at that time, and they have not been discovered since, for the simple reason that they do not and did not exist. They were only the wishful thinking of a committed materialist.

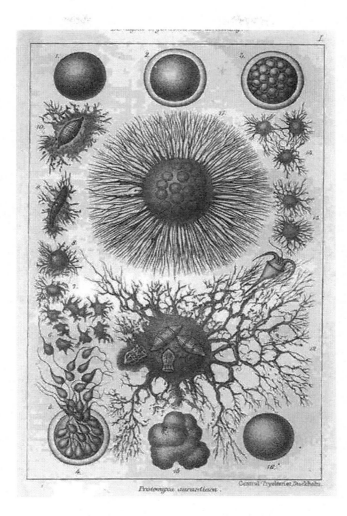

FIGURE 1.2—Imaginary reproductive cycle of Monera drawn by Ernst Haeckel.

In the same era, the Englishman Thomas Huxley gave a talk at the Royal Geographical Society. There he told about a jelly-like matter found on the bottom of the sea. He named it *Bathybius haeckeli*, in honor of Haeckel, his fellow Darwin disciple, and suggested that a slimy layer of the stuff covered perhaps hundreds of square miles on the sea bottom. He proclaimed it the missing link between inorganic matter and organic life. In reality it was only a precipitate that was formed when alcohol was added to seawater.[4]

Darwin contributed to this tradition of imaginative origin-of-life storytelling in an 1871 letter to Joseph Hooker:

> It is often said that all the conditions for the first production of a living organism are now present, which could ever have been present.—But if (& oh what a big if) we could conceive in some warm little pond with all sorts of ammonia & phosphoric salts,—light, heat, electricity &c present, that a protein compound was chemically formed, ready to undergo still more complex changes, at the present day such matter would be instantly devoured, or absorbed, which would not have been the case before living creatures were formed.[5]

Met by such imaginative yarn-spinning it is surely reasonable to ask, what does experimental science actually tell us about the origin of life, fanciful storytelling aside? Part of the answer is that, for a time, experimental science seemed to offer tentative support for the idea that life could emerge spontaneously from very humble source material. The ancient Chinese found evidence that aphids could spontaneously generate from bamboo. Documents from ancient India reference the spontaneous generation of flies from dirt. And the Babylonians concluded that canal mud could generate worms. No less a thinker than Aristotle concurred, seeing no reason to doubt these ancient testimonials.

Later, in the Renaissance, the Flemish chemist and physician Jan van Helmont wrote instructions about how to get mice to emerge from pots containing moist seeds and dirty rags.[6]

But how all this could be remained a mystery. The discovery of microorganisms began to cast doubt on the idea that life regularly and easily sprang from non-life. And eventually the French Academy promised a reward to the one who could solve the puzzle. Louis Pasteur got the prize after showing with an ingenious experiment that living organisms—and specifically in his case, microorganisms—do not form spontaneously. Experiments in the decades that followed confirmed his findings. It was soon conventional wisdom: In the normal course of things, only life begets life.[7]

The hope of finding experimental evidence for the spontaneous formation of life has, however, not been abandoned. It was now clear that life from non-life is not part of the usual course of things, but perhaps it did belong to the realm of the unusual and long ago, and perhaps this possibility could be demonstrated in the lab. The biochemist Alexander Oparin's 1924 Russian-language work *The Origin of Life* offered a partially testable hypothesis for how this might have happened. And John Haldane, apparently unaware of Oparin's Russian-language work, offered a similar proposal in English in 1929. A generation later, in 1953, Stanley Miller put their ideas to the test.

A picture of Miller's equipment (see Figure 1.3) has been featured in practically every biology textbook since then. The public has been led to believe that thanks to Miller's experiment, the problem of the origin of life has largely been solved, at least in broad outline. The 1960 declaration of famous paleontologist George Gaylord Simpson in the journal *Science* is representative. "The consensus is that life did arise naturally from the nonliving and that even the first living things were not specially created," he wrote. "The conclusion has, indeed, really become inescapable, for the first steps in that process have already been repeated in several laboratories."[8]

But here too I came to see that what the school textbooks and the cheerleaders for scientific materialism claimed in public was quite different from what the scientific specialists were saying among themselves. From time to time I would encounter an article or book by some respected scientist in the field expressing the discouraging lack of progress. These confessions came not weeks or months after Miller's experiment but years and decades later.

Making Waves in Darwin's Warm Little Pond

BY THE early '70s I was discussing my growing skepticism of evolutionary materialism with various colleagues and scientific acquaintances. Not all

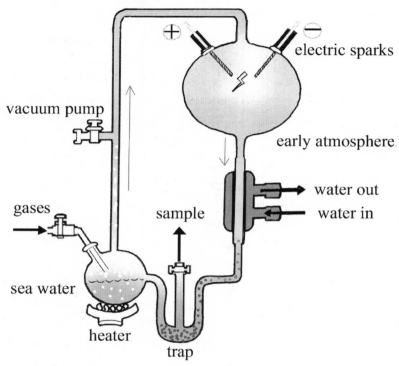

FIGURE 1.3—Schematic presentation of Miller's experimental setup.

of them were close-minded. One of the more open-minded ones worked in the lab next to mine, a biochemistry graduate studying enzymes. For his thesis project, he was studying how fructose was produced from glucose by the glucose isomerase enzyme. For my project, I was optimizing the production of a yeast metabolite (fructose-1,6-diphosphate). I had previously shared with him some of my doubts about chemical and biological evolution, and one day he walked over to my lab with a biochemistry textbook in his hand. "This book discusses things similar to what you have been saying," he announced.

The book is a seminal work in the field of enzymology, appropriately titled *Enzymes*. My colleague pointed me to the last chapter of the book, focused on the question of the origin of the first enzymes in the distant past. There the authors admitted that the topic is extremely difficult and that all attempts to explain the origin of enzymes had failed:

> The difficulties appear to have been greatly underestimated… difficulties seem to have been increased rather than diminished recently. Unfortunately, progress has not been helped by a strong tendency to make light of these difficulties, or even to ignore them altogether… The problem in fact seems as far from solution as it ever was… We are thus led to an apparently insoluble dilemma… The subject is full of difficulties.[9]

My colleague was not ready to throw his faith in scientific materialism overboard, but he was allowing himself to notice that not all was right in the temple of evolution. For my part, I realized that the enzyme question was only the tip of the iceberg. The more I learned about the origin-of-life question, the more skeptical I was becoming of origin-of-life scenarios based solely on undirected chemical evolutionary processes.

My suspicions were further heightened by some overly sunny descriptions of the origin-of-life field. A case in point is Albert Lehninger's *Biochemistry*,[10] a book studied by many biochemistry and medical students in the '70s. Its last chapter focuses on the origin of life and various hypothetical models for the origin of proteins, nucleic acids, and cell membranes.[11] Lehninger's confident references to origin-of-life experiments sounded all very impressive if you were just coming to the field, but by that point I had learned enough to know better. I also noticed that right alongside Lehninger's confident pronouncements were a sprinkling of words like "may" and "possibly." These qualifiers were hints at an unstated reality: Origin-of-life researchers were casting about in the dark, making mostly wild and unsubstantiated guesses.

The Tale of Chemical Evolution: A Primer

Now LET's take a few pages to look at the main points of chemical evolutionary theory to see why efforts to find a blind and undirected cause for the origin of life have failed. If you find it too technical, no worries. You can get the gist of the chapter without this section, so feel free to skim this foray into the technical or just jump straight down to the subheading "The Cell as City." There we'll summarize some of the key takeaways. Next we'll shift from chemical to biological evolution and tell about some leading mathematicians who crashed the evolutionary party. Then we'll wrap up the chapter by looking at a trick some evolutionists use to shut down those who suggest that intelligent design is a better explanation for biological origins.

The story of the unguided, chemical evolution of the first life has some variations depending on whose version you hear, but its main contours can be summarized as follows:

+ At the time when the chemical constituents of the first life were developing, the Earth had virtually no free oxygen, important since the presence of free oxygen would prevent the formation of compounds essential for the origin of life.

+ Nature invented a way to produce the chemical "letters" of the DNA/RNA alphabet: cytosine, adenine, thymine/uracil, and guanine (C, A, T/U, and G for short)

+ Nature invented a way to make the sugars ribose and deoxyribose.

+ Nature invented a way to combine these sugars, phosphoric acid, and the DNA/RNA alphabet letters (the four nucleobases—C, A, T/U, and G) into long chains.

+ Nature invented a self-replicating molecule—DNA or RNA, and eventually both.

+ Nature invented a method to make twenty distinct amino acids. This is a higher-level alphabet consisting of twenty characters.

+ Nature invented a way to combine these amino acids into sophisticated protein machines.

+ After inventing all this, nature changed the self-replicating molecule into a system in which DNA coded for amino acids and thus for proteins.

    ✦   Finally, nature invented a membrane system that isolated the invented molecules from the environment and metabolism began.

As for the point about the early atmosphere containing no free oxygen, later findings have cast doubt on this,[12] and an atmosphere with oxygen spells double trouble for nature's efforts to generate the building blocks of life. But even setting aside that major problem, the invention stages of the story face major obstacles. All the inventions in the bullet points above somehow occurred in the face of one of nature's basic laws, according to which natural systems, when left alone, tend towards disorder and, in the case of chemical reactions, towards equilibrium. Nature, as it turns out, even has trouble getting there with the help of modern technology, brilliant laboratory scientists, and meticulous chemical engineering.

Goop—And Not Even Alphabet Soup

IN STANLEY Miller's vaunted chemical evolution experiments, what was produced was well short of the hype.[13] The result was a tarry slime with 85% tar, 13% carboxylic acids, and 2% amino acids, from which some of the amino acids present in living organisms could be extracted. Many other compounds, some toxic to life, were also formed in the course of the experiment. Similar experiments have been repeated in different laboratories with the same kind of results. They can be summarized as follows:

    ✦   Living organisms have twenty different kinds of amino acids, a twenty-letter alphabet used to "write" protein and protein machines essential to life. But Miller-style experiments produce many amino acids that are not present in proteins. In essence, these amino acids aren't part of the relevant alphabet for coding life.

    ✦   The side chains of amino acids determine their chemical nature. These may be hydrophobic, neutral, acidic, or basic. None of the amino acids with *basic* side chains (lysine, arginine, and histidine) have been formed in Miller-type experiments, and yet these are crucial for life.

+ In any given experiment, only a few, and at most thirteen, of the twenty amino acids present in proteins have been formed. All twenty are needed for life.

+ The composition of compounds formed in Miller-type experiments differs from that found in living cells. Monofunctional compounds that inhibit polymer formation are oversupplied in Miller-type experiments. To form a chain from molecules, the molecules need to have two "sticky" ends; if they have only one, there is nothing for the next compound to attach to. Miller-type experiments produce far too few molecules with two "sticky" ends.

Anyone with a little knowledge of chemistry will see that such a random mixture of chemicals is far, far removed from life's origin.

Sidney Fox also conducted an oft-cited origin-of-life experiment. He studied the polymerization of pure amino acids in dry conditions at around 170 degrees Celsius. He produced simple polymers that he called proteinoids, known also as thermal polypeptides. These molecules contained chemical bonds not present in the functional three-dimensional proteins of life, and so they have precious little to do with life's origin, or with the biological information essential to life. The information in DNA and in the proteins DNA codes for are mostly non-repeating, complex, functional sequences, akin to computer software code or the letters and words in a novel or instruction manual. The polymers that Fox produced are nothing like this. Stanley Miller and Leslie Orgel emphasized this crucial difference in their 1974 book *The Origin of Life on Earth*. After underscoring the difference, they concluded, "It is deceptive, then, to suggest that thermal polypeptides [proteinoids] are similar to proteins."[14]

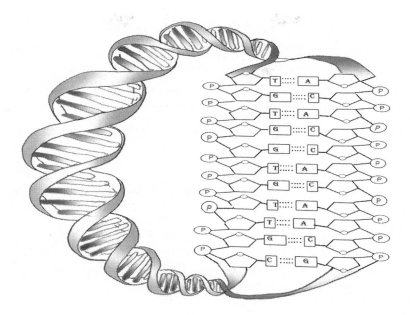

FIGURE 1.4—The DNA double helix, containing tightly packed coded hereditary information. It's so tightly packed that the entire human genome weighs only six picograms. A picogram weighs a mere one trillionth of a gram.

An RNA World of Trouble

THOMAS CECH and Sidney Altman (Nobel Prize, 1989) discovered that *ribo*nucleic acid (RNA) molecules make some life-essential chemical reactions possible. The discovery of "ribozymes" (ribonucleic acid enzymes) that not only store genetic information but also serve as biological catalysts led to a new origin-of-life scenario, the "RNA World" hypothesis. On this view, an RNA world was an important early step on the way from dead chemicals to living cells. However, no one has been able to show how RNA could be formed by random chemical reactions, or survive for long after it did form.

One of RNA's jobs is to take the information in DNA and translate it into protein molecules. Both RNA and DNA are made of three parts. Both contain phosphoric acid and nucleobase. DNA's third part is a sugar called deoxyribose. For RNA, it's a delicate sugar called ribose—indeed, alarmingly delicate given that RNA-World proponents are hoping that RNA might have been present at the origin of the first life. Ribose is one of the most reactive sugars, a quality evident in controlled laboratory conditions, and a hardly helpful one in the hurly-burly of a primordial soup. Ribose reacts easily with proteins. I have studied enzymatic conversions of ribose, and it tends to form chemical bonds with enzymes. That's a problem for unguided chemical evolution scenarios because chemical reactions between ribose and amino acids would destroy any imagined proteins.

No credible route to the formation of ribose in prebiotic conditions is known although it can be engineered from formaldehyde together with several other compounds. Nucleobases can be formed in concentrated cyanide (HCN) solutions, but if we look closer we see that this isn't something to get excited about: Their synthesis is difficult, the obtained concentrations small, and their stabilities low.

Thus, the RNA-World hypothesis is without a real basis in chemistry.

This is a key reason Robert Shapiro rejected the RNA-World idea in his 1986 book *Origins*.[15] Having rejected it, he proffered an alternative explanation. "A more likely alternative for the origin of life is one in which a collection of small organic molecules multiplies their numbers through catalyzed reaction cycles, driven by a flow of available free energy," he wrote. Note, however, that he did not reject the RNA World due to any great confidence in this alternative scenario. "Although a number of possible systems of this type have been discussed," he added, "no experimental demonstration has been made."[16]

Both RNA-first and DNA-first origin-of-life scenarios face a significant challenge. Nucleic acids consist of a series of nucleotide subunits. These nucleotide subunits—made of a sugar + a phosphoric acid + a nucleobase—get connected into long chains to form the nucleic acids RNA or DNA. So nucleotides are essential to form RNA and DNA chains. However, there is no apparent way that nucleotides can be formed in the imagined prebiotic soup. This is a key reason why there are no believable models for the synthesis of RNA or DNA molecules by undirected chemical reactions.

When Two Hands Aren't Better Than One

MANY OF the molecules of life have two energetically equal forms. They resemble each other like a left and a right hand. When a molecule has a form such that it could theoretically have a right and a left-hand form, it's said to possess *chirality*. A pair of three-inch-diameter circles do not possess chirality. They are, after all, interchangeable and identical. The human hand does possess chirality. A left hand isn't simply a right hand that could be flipped and stuck on the left arm. Imagine it and you'll see that they aren't interchangeable that way, even if there were a lurid Frankenstein doctor capable of performing the surgery. In living organisms, there are molecules that possess chirality, but usually only one of these two forms is present, such as D- and not L-ribose in RNA. It would be a bit like all humans only possessing left hands. We wouldn't have right hands, but we could imagine what a right hand would look like and how it would differ from a left hand.

Yet the analogy to human hands breaks down. In humans, it obviously works better to have a right and a left hand, rather than only left hands or only right hands. For chiral molecules, the important thing is this: To be functional, proteins and nucleic acids must be formed from all one or the other. The molecules must have the correct chirality. To use the language of hands again, in a particular case you'll need all left-handed molecules for things to fit together and for each to function as a building block.

That's a challenge for unguided origin-of-life scenarios. In chemical reactions both chiral forms are produced in equal amounts. That is, you get about equal portions of left-handed and right-handed molecules. There is no known way to produce by random chemistry only one of these forms. It's like flipping a normal, two-sided coin a thousand times. If it's a fair coin and fair tosses, you will get close to half and half heads and tails. There is no random process that would consistently yield all heads or all tails.

When an organism dies, we see a kind of reversion to the mean. The amino acids of proteins in living things are all left-handed. After death, they slide back slowly toward an even mix of right and left-handed. To use the technical terminology, the molecules start to *racemize*. In essence, without life already present, the law of entropy takes over and disorder reigns. So how do you suspend the law of entropy in order to generate the first life by blind processes, since you need life in the first place to suspend the law of entropy?

To sum up: Miller's origin-of-life experiment formed a few amino acids—but as a racemic mixture. That is, the amino acids were a roughly equal mix of left-handed and right-handed components. That's because, again, given over to its own devices, chemical ingredients mixing around together tend strongly toward a roughly even mix of left and right, much as you would expect to get a roughly even number of heads and tails if you flipped a coin a thousand times. The trick that evolution needs to perform is to get those life-essential amino acids to form all with one orientation (either all left or all right) and to do so through a blind, natural process. Miller's experiment accomplished no such thing.

Self-Organizers (or No, Aliens!) to the Rescue

Manfred Eigen and Ilya Prigogine were awarded a Nobel Prize for their studies on physical systems that are far from equilibrium. Their ideas have been applied to the theoretical pondering of life's origin. Stuart Kauffman and the Santa Fe group have suggested self-organization as a solution to life's origin.[17] They talk of chemicals spontaneously forming into cells. Eigen speaks of hypercycles that formed the first cells. Attempts to prove these ideas experimentally have failed. The self-organizational models are algorithmic and repetitive in nature, and as the late mathematician and doctor of medicine Marcel-Paul Schützenberger argued, such patterns are information-starved compared to the living world, and non-functional.[18] According to him, no algorithm can describe the complexity of living organisms. Try to imagine a mathematical algorithm spitting out a great novel from scratch, and you have some sense of what he was getting at.

How bleak are the prospects for explaining the origin of life without reference to a creative intelligence? Bleak enough that some leading origin-of-life researchers have taken to evoking alien life to get around the challenge. Swedish Nobel-Prize winning chemist Svante Arrhenius suggested that life's seeds were originated somewhere else in the universe and then somehow made their way to Earth. Francis Crick is probably the best-known supporter of this idea. After realizing the enormous problems of chemical evolution, he tried to find an escape in this direction. But the solution doesn't solve the problem. It just moves it, and creates new problems into the bargain. The possibilities of life surviving in interstellar space for long ages have been studied using bacteria. Studies by Klaus Dose and Anke Klein show that radiation damages bacterial spores. Thus, there are clear limits for the time and distances available, limits far short of the many trillions of miles that separate our solar system from other stars.[19]

The Cell as City... Bustling with Information

GIVEN ALL the above talk about nucleic acids, proteins and such, it's important to underscore that a living cell is much more than just nucleic acids and proteins. It has the sophistication of a factory or city. Whole books could be written about its complexity (and have been), and unplumbed mysteries would remain. Here we'll get only a brief taste of that complexity, and look at just a few of its essential elements.

A complex cell membrane is necessary to separate the content of the cell from the environment. It is always formed from pre-existing membrane. It separates the intracellular reactions from the environment. A cover that separates the complex reaction pathways would mean their isolation from the outside world and the end of the cell if there were not specific transport systems through the membrane. But without a membrane, the complex reaction pathways would stand no more chance of surviving and succeeding in the primordial soup than a house of cards in a storm. So, the origin of the membrane must be associated very closely with the formation of specific transport systems. The membrane, in other words, is likely an essential part of any irreducibly complex first life able to survive and reproduce. A sort of pre-cell without a membrane would thus be a dead end—dead on arrival.

What else does a "simple" single-celled organism need to survive and function? In 1995 a research team led by Craig Venter published the complete genome sequence of *Mycoplasma genitalium*. This organism is a parasite. It has one of the smallest genomes, with about 480 protein-coding genes. Life doesn't get much simpler than this. But the gap between this organism and the experimental results of chemical evolution is enormous.[20]

When the so-called archaebacteria were discovered, some scientists speculated that these organisms could offer a nice model for the first systems that chemical evolution had produced. But studies of these organisms have revealed fascinating metabolic systems that are far from simple. Archaebacteria are actually metabolic masters.

An essential property of all life, including archaebacteria, is information. There's the information written using DNA's four-letter alphabet, and there's the information in the proteins built using the instructions from DNA. But the chemical structure of DNA does not explain its code—that is, the rules that cells follow in translating the information in DNA into all functional proteins. Nor does it explain the software written by it. The chemical structure doesn't explain it any better than the chemical composition of ink and paper explains the information in a book, or the language and grammatical rules used to record the book's message.

Where did the genetic code come from? How could it change and remain functional at each evolutionary step? We find variations in some organisms and in mitochondria, so if the genetic code evolved into its different forms, it must have changed. In the Oxford University Press book *Evolution of the Genetic Code*, Syozo Osawa concludes that we can only make observations and guesses about the origin of genetic language. As he concedes in the book's final sentence, "The origin and early evolution of the genetic code, and so the origin of life itself, still pose an enormous challenge."[21]

The reality is that biological information remains an enigma for those committed to purely materialistic origins scenarios. We have no scientific knowledge supporting the notion of a mindless origin for this essential feature of life. And we have good reasons to conclude that biological information, and the language it is written in, instead have their origin in the work of a creative intelligence. I explore this question more in a later chapter.

Now someone might object that scientists have in recent years uncovered evidence that life may have appeared on Earth soon after the conditions were right, so hey, how hard could it be? But the timing says almost nothing about how the life arose. Was it a mindless cause? Was it intelligent design? A clue as to the when does not answer the who and the how.

The one thing that this finding does do is make more trouble for scenarios depending on chance. If you play a state lottery for only a short time, you have a worse chance of winning the jackpot than if you play it for a long time. It's the same with the chance origin of life.

That being said, we need to be careful with analogies to state lotteries. There is no human lottery that remotely approaches the long odds involved in the chance origin of the first life. Based on our current knowledge of what the origin of life would require, it appears that a trillion years times a trillion years wouldn't have been long enough. The simplest self-reproducing organism is so insanely complex that the amount of time needed for luck to have a fighting chance vastly exceeds the age of the whole universe, and now we have a window both much shorter than that and much shorter than previously believed.

The "official" view remains that life appeared spontaneously, not long after the conditions were right, with no need for intelligent design. But there is no evidence for this view, and laboratory work on the issue has only made matters worse. Fred Hoyle provided a good summary of the matter. "If there were some deep principle which drove organic systems towards living systems, the operation of the principle should be demonstrable in a test tube in half a morning," he wrote. "Needless to say, no such demonstration has ever been given. Nothing happens when organic materials are subject to the usual prescriptions of showers of electrical sparks or drenched in ultraviolet light, except the eventual production of a tarry sludge."[22]

Hoyle wrote that more than thirty years ago, and laboratory work in the intervening three decades has only corroborated those findings. We have no evidence for an unguided origin of life, and mounting experimental evidence against it. The idea remains sheer speculation.

James Tour is a leading origin-of-life researcher with over 630 research publications and over 120 patents. He was inducted into the National Academy of Inventors in 2015, listed in "The World's Most Influential Scientific Minds" by Thomson Reuters in 2014, and named "Scientist of the Year" by *R&D Magazine*. Here is how he recently described the state of the field:

> We have no idea how the molecules that compose living systems could have been devised such that they would work in concert to fulfill biology's functions. We have no idea how the basic set of molecules, carbohydrates, nucleic acids, lipids and proteins were made and how they could have coupled in proper sequences, and then transformed into the ordered assemblies until there was the construction of a complex biological system, and eventually to that first cell. Nobody has any idea on how this was done when using our commonly understood mechanisms of chemical science. Those that say that they understand are generally wholly uninformed regarding chemical synthesis. Those that say, "Oh this is well worked out," they know nothing—*nothing*—about chemical synthesis—*nothing*.
>
> ... From a synthetic chemical perspective, neither I nor any of my colleagues can fathom a prebiotic molecular route to construction of a complex system. We cannot even figure out the prebiotic routes to the basic building blocks of life: carbohydrates, nucleic acids, lipids,

and proteins. Chemists are collectively bewildered. Hence I say that no chemist understands prebiotic synthesis of the requisite building blocks, let alone assembly into a complex system.

That's how clueless we are. I have asked all of my colleagues—National Academy members, Nobel Prize winners—I sit with them in offices. Nobody understands this. So if your professors say it's all worked out, if your teachers say it's all worked out, they don't know what they're talking about.[23]

Despite all this, Stanley Miller's experiment is still presented in text-books as if it all but sealed the deal for a naturalistic origin of life. NASA is still searching for marks of life on nearby planets, fueled by the belief that life should spring up relatively easily given the right conditions. And the uninformed public continues to be told that life is nothing more than complex matter. There seems to be only one explanation for this stubborn refusal to register all of the contrary evidence. We are dealing with a conviction deeply rooted in a worldview.

This explains how a physics professor in a major Finnish newspaper can say the following with a straight face: "The question of the origin of life from the viewpoint of nanotechnology is almost without content. There is no qualitative difference between life and non-life."[24]

FIGURE 1.5—Worldview can shape how people interpret an observation and what they consider important. In this picture, some see a beautiful woman, while others see an old lady. Similarly, for some, the biochemical mechanisms of a cell appear designed, while others see them as a product of chance and selection.

Take that assertion in for a moment. This is how one of the most cited Finnish naturalists proclaims unconsciously his own faith. To avoid the overwhelming problem facing the materialistic theories for the origin of life, he simply pretends that the line between life and non-life is largely meaningless. A worldview that must dispense with something as basic and undeniable as the distinction between life and non-life is a worldview in crisis, even if its followers have a marvelous capacity for pretending otherwise.

Mathematicians Crash the Evolution Party

ANOTHER EYE-OPENER for me was the book *Man's Origin, Man's Destiny* by the late professor and organic chemist A. E. Wilder-Smith.[25] The book analyzed the probability (or improbability) of chemical reactions creating information and machines, and it referred to a discussion between mathematicians in the summer of 1965 in Switzerland, which led to a famous meeting of mathematicians and biologists at the Wistar Institute in 1966.[26] The mathematicians expressed doubts about the creative powers of blind evolution, and specifically argued that evolution via the neo-Darwinian mechanism was simply too improbable mathematically to be plausible.

Among the critics present was Professor Murray Eden from MIT. Another was Marcel Schützenberger, who later became a professor in the University of Paris and a member of French Academy of Sciences. He was till the end of his life very critical of evolutionary theory, as evidenced by his last interview.[27] When Schützenberger explained the results of his simulation experiments at the Wistar meeting, evolutionary biologist C. H. Waddington shouted at him, "We are not interested in your computers." – "But I am," Schützenberger responded.

Molecular biologist Douglas Axe was also interested in the problems raised at Wistar, and, some four decades later, he pursued a series of experiments at labs in and around the University of Cambridge to shed light on the challenges raised at Wistar.

In software programs and in human languages, non-functional sequences are vastly more common than functional sequences. This is a key reason why random changes to a book's text, or (even more so) to a software program, degrade its meaning or function, and it is why attempts to evolve meaningful sentences or functional software code through a truly neo-Darwinian process have failed.[28] But maybe genetic information is different. Maybe the proportion of functional to dysfunctional sequences is much higher for genetic information than for books and software code. In search of an answer, Axe studied a small unit of biological function, proteins—each one roughly equivalent to a meaningful sentence in a book. Axe wanted to see how rare functional proteins of a certain length were among all possible sequences of that length.

Recall that the four-character DNA alphabet helps code for the twenty-character amino-acid alphabet, and the twenty-character amino acid alphabet helps code for the many different kinds of proteins. A particular sequence of amino acid letters, in other words, serve as building instructions for a particular kind of protein. A protein with function A will have a very different sequence of amino-acid letters than a protein with function B, just as two software programs, each with a very distinct function, will have very distinct lines of code one from the other, even if there is a bit of overlap here or there.

Why look at humble proteins instead of something more biologically complicated? The wide array of different proteins we find in living things serves many essential and distinct roles for cells and cellular machines. For a life-form to evolve into a new and highly distinct life-form, existing proteins must evolve into new and very distinct proteins. If the neo-Darwinian mechanism can't evolve new proteins, it can't evolve anything new in the biosphere. The evolutionary process is stuck in the mud.

To return to Axe's story, he chose to study proteins (enzyme proteins in this case) because they have sequence specificity. They also have a measurable chemical function. That meant Axe could vary the combinations of amino acid sequences, and measure the resulting proteins' chemical activity to see if they still worked as functional enzymes.

So what did Axe discover? If the functional proteins were too rare, then it would mean biological information is like books or software code: You can't evolve fundamentally new and functional information through a blind process because there is just too much non-functional gibberish to wade through. Axe looked at proteins of modest length (150 residues) and published his results in the *Journal of Molecular Biology*.[29] He found that the ratio of functional proteins to non-functional gibberish was 1 in 10^{74}. He found that the odds of getting a protein with a particular function was 1 in 10^{77}. That's one protein capable of carrying out that function for every 100,000,000,000,000,000,000,000,000,00 0,000,000,000,000,000,000,000,000,000,000,000,000,000,000, 000 dead-on-arrival wannabe proteins.

Calling that a needle in a haystack is to flatter haystacks. The number of atoms in all of planet Earth is estimated to be around 10^{50}—a huge number but one dwarfed by 10^{77}. The latter number is a billion times a billion times a billion times bigger.

Axe also showed that if all the life on Earth for billions of years was busily searching via random mutation for even one new protein in that cosmic-sized ocean of non-functional protein gibberish, it couldn't find it. And, of course, a new life-form requires not one but many hundreds of new protein types along with lots of tricky epigenetic information, so we're talking about an absolutely essential but far from sufficient condition for blindly evolving new biological form and function. Axe's findings, in other words, corroborated what the mathematicians at Wistar had suspected and argued: neo-Darwinism had a major math problem.

Something Rotten in the State of Darwin

AXE'S FINDINGS were still decades in the future, but already by the 1970s I had begun to find the case for blind evolution increasingly implausible. Wilder-Smith understood that the central role of DNA is to carry information, not just to function as a complex chemical compound, and he saw this distinction as decisive. He used simple but instructive examples to clarify his argument. So for instance, paper and ink, he noted, do not write a book. Another example he gave: a can full of sardines. The contents of the can have all the necessary building blocks and even information-rich polymers. The can is also thermodynamically open—energy can be imported and exported. And the atmosphere inside the can is reducing. In spite of these favorable conditions, no ordering towards life occurs—just the opposite: a process leading to energy minimum and degradation occurs.

Wilder-Smith realized that a biological cell contains various molecular machines. He described chlorophyll as a metabolic engine that converts the Sun's energy to chemical energy. Without this engine, there would be no life on Earth. The Sun would only dry everything. He used the following illustration: You can pour gasoline on a car and light it, but the car moves nowhere. The gasoline must burn in an engine that converts the liberated energy to propulsive kinetic energy.

Yes, evolutionists have developed responses to these and other arguments described above, but again and again as I have followed this debate over the years, I have found their counterarguments inadequate. By the middle of the 1970s my doubts had become a conviction: Scientists have no materialist explanation for the origin and complexity of life. The confident bluffing of the dogmatic materialists notwithstanding, they weren't even close. Experimental science, I concluded, seemed to point in a different direction.

In the following decade, I found further encouragement for my heretical thinking, and from an unlikely source: a Nobel-Prize winning geneticist. In 1987 I met Professor Werner Arber during a symposium where I was given the Latsis Prize, given to a young scientist under 40 working at the Swiss Federal Institute of Technology (ETH). Arber had won a Noble Prize in 1978 with two American scientists for their discovery of important tools used in genetic engineering. Arber was not sold on the idea of a purely naturalistic origin of life and considered an intelligent cause a satisfactory explanation. Here is how he put it in an interview:

Although a biologist, I must confess that I do not understand how life came about.... I consider that life only starts at the level of a functional cell. The most primitive cells may require at least several hundred different specific biological macro-molecules. How such already quite complex structures may have come together, remains a mystery to me. The possibility of the existence of a Creator, of God, represents to me a satisfactory solution to the problem.[30]

In my conversation with him after the symposium, I was impressed that he came to congratulate me for my achievement and that he remarked positively about my briefly thanking God for blessing my research work at ETH. I only later learned about his views in this field. It made an impression on me. Here was an experimental biologist of the first order who was open to the idea that a designing intelligence played a role in the origin of life.

And clearly that openness had not crippled him as a scientist. Why should it? The question of when, how, and where such a creative intelligence might work remained an open question for him, so continued investigation and experiment remained essential parts of his exploratory toolkit. He wasn't replacing one dogma with another. He was refusing to

FIGURE 1.6—With my daughters at Latsis-Prize ceremony in 1987.

be dogmatic at all. I liked the sound of that: Set aside a rigid allegiance to matter-only explanations and simply follow the evidence.

The Materialism of the Gaps

A COMMON response to the utter failure to discover any way that the first live cell could have originated apart from intelligent design is to counsel patience. "We just have to wait patiently till a purely naturalistic answer emerges," the argument goes. "Let's not grow impatient and start stuffing God or an 'intelligent designer' into the gap of our knowledge just because we cannot find an answer right away." That response had once seemed unanswerably wise to me, but no longer.

Consider an analogy. Imagine you have gone to visit the circular pattern of great stones on the Salisbury Plain in England known as Stonehenge. As you walk around it, admiring its geometrical precision and its relation to certain astronomically significant patterns, you comment to your traveling companions that clearly whoever designed and built Stonehenge was no dummy. At that moment, a stranger beside you turns to your little group and says, "Look here now, don't lose your heads and start invoking ancient Druids or Beaker folk or mysterious Leprechauns with a rage for hauling massive stone pillars miles and miles cross country to this precise spot. The origin of Stonehenge surely has some purely material explanation. We just have to be patient enough to keep searching for it."

Those listening to the stranger would be justified in thinking, *what an odd way to approach the question.* Indeed, the fellow's thinking would only be reasonable if somehow we already knew that the pattern in question had a purely material cause for its origin and were just trying to work out some of the details. But if we have reason to suspect that it was designed, or even if the cause for the origin of Stonehenge remains uncertain, then insisting we all hold out for a satisfactory materialistic explanation is only so much question-begging.

Of course, this illustration is only that—an illustration. The simplest living organism is vastly more sophisticated than the circular arrangement called Stonehenge. And the first living cell obviously was not the work of human designers. But the basic point of the illustration holds: If something possesses a common hallmark of intelligent design—namely the sophisticated arrangement of parts that accomplishes some striking purpose—one cannot rationally refute the design hypothesis simply by ruling that explanation out of court from the outset.

That much was now clear to me, and having realized it, there was no going back to the old, question-begging way of viewing things in biology.

2. Fossilized Materialism

W E ALL HAVE A TENDENCY TO AVOID KNOWLEDGE AND OPINIONS that threaten our position and worldview. We are usually more interested in personal peace than in truth. We tend to put off uneasy thoughts. And scientists are no different from other people. A scientific investigator's framework or paradigm often hinders him or her from fairly considering other views. The temptation is to consider a contrary paradigm to be not just in error but heretical.

I learned of one particularly striking example of this from a student. He told me that he had offered a Finnish version of an excellent German textbook, *Evolution: A Critical Textbook* (*Evolution—Ein Kritisches Lehrbuch*), by Reinhard Junker and Siegfried Scherer,[1] to a genetics professor. According to the student, the professor told him he was not willing to read heretical books. By heretical he meant views that contradicted his own.

I tried to correspond with the professor, but it ended quickly because my views—according to him—were not scientific. By "scientific" he meant materialistic. He meant that I didn't hew to the rule of methodological materialism: No theories inconsistent with materialism/atheism are allowed. When I asked him how he explained the origin of a bacterial electromotor, he said evolution takes care of that easily. No details. Just boundless faith in the creative powers of evolution.

I encountered a similar close-mindedness after I published an earlier version of this book in Finnish. I gave it to several professors and colleagues. When I later asked them what they thought about it, the usual answer was that they had not yet had time to read it.

Since the 1970s the Finnish academic world has been shaken repeatedly by evidence and arguments critical of evolutionary theory. Well-known scientists, philosophers, theologians, and bishops of the Lutheran church have been quick to present a united front against this onslaught, reassuring the public that the theory of evolution has been convincingly proved and is not in conflict with a correct interpretation of the Bible, so there is no need to pay any attention to the naysayers. They go further and warn that these naysayers only undermine the reputation of Finland's universities and the authority of the church, since both have thrown in their lot with modern Darwinism.

French evolutionist Thomas Lepeltier, no fan of intelligent design theory, argues that such reactions move us toward a society where all challenging of scientific theories will cease, an outcome that would be terrible for science, he warns.[2] Science, after all, thrives on open inquiry, critical analysis, and debate.

Already, the materialistic paradigm controls most of the discussions around the origin of life and the origin of species. Michael Ruse, a philosopher of science, thinks that evolution is true but admits that for many it has become a religion. "Evolution is promoted by its practitioners as more than mere science," he writes. "Evolution is promulgated as an ideology, a secular religion—a full-fledged alternative to Christianity... Evolution is a religion. This was true of evolution in the beginning, and it is true of evolution still today."[3]

This religion controls origins research so tightly that, according to cell biologist Franklin Harold, "We should reject, as a matter of principle, the substitution of intelligent design for the dialogue of chance and necessity" even though "we must concede that there are presently no detailed Darwinian accounts of the evolution of any biochemical or cellular system, only a variety of wishful speculations."[4]

What is the principle Harold is talking about? He does not spell it out, but clearly he is talking about methodological materialism. The science community accepts it as self-evident, often without recognizing its dogmatic nature. The late German physicist and philosopher Carl

Friedrich von Weizsäcker was fair enough to admit this. "It is not by its conclusions, but by its methodological starting point that modern science excludes direct creation," he wrote. "Our methodology would not be honest if this fact were denied."[5]

For over forty years I have had numerous discussions both within and without the science community concerning the origin of life and the origin of species. Practically all of the hundreds of scientists I know admit in private, confidential discussions that science does not have a clue where genetic language, proteins, cell membranes, metabolic pathways, cell control systems, and the basic body plans of organisms came from, just as Franklin Harold admitted in *The Way of the Cell*. In spite of that, their only acceptable creation story is materialistic evolution. "A key symptom of ideological thinking is the explanation that... cannot be tested," says Stanford University physicist and Nobel laureate professor Robert B. Laughlin. He calls "such logical dead ends antitheories" and says that "evolution by natural selection... has lately come to function more as an antitheory, called upon to cover up embarrassing experimental shortcomings and legitimize findings that are at best questionable and at worst not even wrong."[6]

Methodological materialism poses as "the scientific method"—empirical, neutral, disinterested. But this isn't the case. It is not a neutral way to observe the world. It dogmatically limits possible answers. The possibility that life has been designed is deemed out of the question. In 1999, S. C. Todd put it plainly in the journal *Nature*: "Even if all the data point to an intelligent designer, such an hypothesis is excluded from science because it is not naturalistic."[7]

The Fossil Pattern

THE SCIENCE literature, of course, isn't in the habit of conceding that "all the data point to an intelligent designer." Instead, the literature is replete with claims that blind, unguided evolution is an established fact. When I ask how we can know evolution is true, the answer often involves some reference to the fossil record. I perused various high school biology textbooks of the last fifty years and without exception they all had the same stories about horse fossils and horse evolution, the archaeopteryx fossil purporting to prove dinosaur-to-bird evolution, and fossils of extinct primates said to be on the evolutionary path to humanity. I used to think that such fossils are the best evidence for evolution, but then I started to investigate the evidence.

Each of these commonly cited evidences for modern evolutionary theory has major problems that become apparent as soon as you push past the introductory textbook salesmanship and really scrutinize them. There you find that even mainstream evolutionists admit there are big problems with these icons. Jonathan Wells details all this in his 2017 book *Zombie Science: More Icons of Evolution*.[8]

And the problem is bigger than a handful of fossil failures. It's the very pattern of the fossil record. New animal body plans appear abruptly in the fossil record and then persist largely unchanged until they disappear.

The Cambrian explosion is only the most dramatic example of this pattern. Evolution skeptics didn't give it that name, by the way. The *Cambrian explosion* is just a common term paleontologists use to describe this

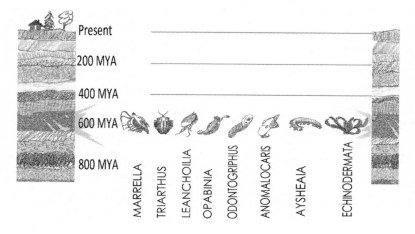

FIGURE 2.1—A majority of animal phyla first appeared in what is known as the Cambrian Explosion.

period in the history of life, and this despite the fact that the apparent explosiveness of the event is a major headache for evolutionists.

Stephen Meyer explores this in his 2013 book *Darwin's Doubt*. There he points out that Darwin himself saw the sudden appearance of dozens of new animal forms in the Cambrian period as a major problem for his theory. He hoped that future discoveries would come to his rescue, but a century and a half of further investigation has only made the problem worse.

An early stage in this bad-to-worse progression occurred early in the twentieth century. In 1909 Charles Walcott, who oversaw the Smithsonian Institution, made an important fossil discovery in the Canadian Rockies. Named the Burgess Shale, it contained a trove of astonishingly well-preserved Cambrian-era fossils. The discovery suggested that the Cambrian explosion had produced an even greater variety of new animal body plans than previously believed.

FIGURE 2.2—In 1909 Charles Doolittle Walcott, exploring in the Canadian Rockies, made one of the most important Cambrian fossil discoveries of the last century. Among these Burgess Shale fossils were many previously unknown animal body structures. The discovery revealed huge gaps in the purported evolutionary chain.

Then, at the other end of the twentieth century, in 1995, J. Y. Chen discovered another Cambrian-era fossil site in China, one that suggested the Cambrian explosion was even shorter, and more diverse, than previously believed.

Other advances in paleontology in the twentieth century made it increasingly implausible that the Cambrian explosion was merely apparent—that is, just an artifact of an incomplete fossil record.

So what's the problem with a relatively sudden appearance of numerous new phyla in the fossil record? Why did Darwin, and now many contemporary evolutionists, consider the Cambrian explosion a problem for Darwinism? If Darwin is correct, then every new animal body plan evolved from earlier and distinct biological forms through a series of tiny evolutionary steps spread out over many hundreds of millions of years. (More below on why the steps need to be tiny ones.) Even with an incomplete fossil record we should find abundant evidence of that gradual branching process of one or a few original forms leading to this menagerie of new body plans. Instead we get no viable precursors and then, suddenly, dozens of new and highly distinct body plans in the aptly named Cambrian explosion.

An important early skeptic of Darwin's theory was paleontologist Louis Agassiz of Harvard. He knew the fossil material better than anyone else in his day. With the help of sailors and missionaries, he collected tens of thousands of fossil samples and identified over ten thousand new species. Thanks significantly to his efforts, Harvard was at that time the most important natural history museum in the world. Darwin himself praised Agassiz in his private correspondence and hoped to win him to his view about the history of life.[9] But Agassiz remained skeptical to the end.

Agassiz offered the pattern of the fossil record as a major reason for his skepticism. Why were the key fossils missing, the very fossils that might prove Darwin's theory correct? They were missing at several crucial stages in the history of life, and most dramatically in the period leading up to the Cambrian explosion. If it were just a question of an incomplete fossil record, you still wouldn't get the specific pattern of abrupt appearance and long stasis, Agassiz insisted.[10]

Darwinism's Missing Marbles

CONSIDER AN analogy. You and your friends are driven to a giant field covered a foot deep in marbles. It's dawn and still too dark to see colors, but your host informs you the marbles there in the field come in a myriad of colors, so various that if a sample of each color type in the field were placed end to end with samples of all the other color types, they would form an exquisitely high definition rainbow progression across the visible light spectrum, hardly distinguishable in its subtle progression from the visible light spectrum itself. This progression from one color to another would be so smooth, in fact, that only a very close inspection would reveal any change at all from one marble color to the next one beside it.

At this point you and your friends are each paid a nice sum to make the day's labor worth your while and are released to wander blindfolded through the field selecting marbles at random and bringing them back to the base camp.

Each of you make several trips back to the base camp loaded with bags of marbles, but somehow all of you keep returning to camp with all primary-colored marbles, again and again and again. To speed the work, you are given wheelbarrows. After a while the base camp is sporting three big marble piles—one red, one blue, and one yellow. Eventually one or two of you come across an orange marble, an occasional purple marble, a few green marbles, and a couple of grayish ones with maybe a hint of blue if you hold it in the light just right. But there's never anything remotely approaching the perfect rainbow of variation you were promised.

You don't want to be rude, but finally you mention this to your host.

"Well," your host explains, "it turns out that only a small fraction of the original marbles are still around—a very, very tiny fraction."

"Oh," you say, thinking you're starting to understand. "So whoever carted the rest of them off went after the transitional colors, I guess.

Maybe they wanted to have all primary colors in the field, and they just overlooked a small number of the marbles that weren't red, yellow or blue. It's pretty amazing they missed as few as they did. They must have had a real mania for leaving behind only primary-colored marbles."

Your host looks aghast. "*Choose?* It wasn't by design. It was by utter happenstance. What are you suggesting!"

The host's response is so bizarre that you decide he must have misunderstood you, so you try one more time to make your meaning clear without raising the temperature of the conversation. "What I'm wondering is, if the field was once a rainbow of thousands of different-colored marbles, and if most of those marbles were carted off more or less at random, well then…" Here you gesture at the field, casting about for some way to make clear to the host a point that should be clear to any objective mind. But before you can finish your comment, the host cuts in. "The orange and purple marbles!" he cries. "Can't you SEE? The missing *links!*"

Now, of course, the history of life is more wide-ranging and variegated than any rainbow, but the marble analogy does accurately capture the problem with the attempt to pass off a few extinct animal body plans and species as proof of gradual Darwinian evolution. These attempts just don't wash because, on Darwinian grounds, we should expect to find millions upon millions of distinct transitional forms, even if we have fossilized remains of only a tiny, tiny fraction of various animal forms that have ever lived on Earth. The pattern of abrupt appearance is one that fits with an intelligent design hypothesis. Intelligence can proceed by great leaps. The pattern doesn't at all fit the neo-Darwinian model.

American paleontologist Robert L. Carroll is in the evolution camp, but he recognizes the problem. "The most striking features of large-scale evolution are the extremely rapid divergence of lineages near the time of their origin, followed by long periods in which basic body plans and ways of life are retained," he writes. "What is missing are the many intermediate forms hypothesized by Darwin."[11]

Paleontologists Stephen Jay Gould and Niles Eldredge, like Carroll, have frankly admitted the problem with the fossil record. In a 1972 paper they offered an alternative solution, a revised evolutionary model called *punctuated equilibrium*.[12] According to this idea, evolution moves by relatively quick bursts followed by long periods of stasis. This would mean that we should expect to find far fewer transitional forms in the fossil record than if evolution always moved at the same plodding pace.

This proposed solution, however, has its own weaknesses—problems acute enough that many mainstream evolutionary biologists continue to view it skeptically. First, even the quick bursts proposed by Gould's model require many millions of years to get major new forms. That's because, as Gould and Eldredge themselves have conceded, natural selection working on beneficial genetic mutations still must do the primary creative work, and that can only happen one small step at a time.

Why not big jumps? Because big random mutations don't improve fitness. They maim and kill. This is well established experimentally, and the reasons for it are discernible from an analysis of engineering constraints at the molecular biological level. (See Chapter 16 of *Darwin's Doubt* by Stephen Meyer for why evo-devo and other patches offer no escape from this problem.) Evolutionist Richard Dawkins used the illustration of what he called Mount Improbable. The front of the mountain is a huge cliff. There is no way to scale it, just as there is no way for big mutations to generate fundamentally new biological form and function fit enough to get passed along in the evolutionary game of life. Unlike in the comic books, big mutations produce dysfunctional, sterile, and even stillborn offspring, not genetic superstars. But, Dawkins says, the backside of Mount Improbable is a gradual slope to the top. This gradual slope represents the neo-Darwinian process of natural selection preserving and passing along various useful micro-mutations over hundreds of millions of years. This, Dawkins and most other evolutionists insist, is the only possible way up Mount Improbable. It's the only way, in other words, for blind evolution to produce novel organs and novel body plans.

But in refuting punctuated equilibrium, Dawkins has jumped out of the frying pan and back into the fire. On the traditional evolutionary model, we should expect to see extremely gradual and incremental changes in the fossil record from one form to the next in the tree of life, but the fossil record fails to cooperate, whether in the Cambrian explosion, or in the emergence of birds and land animals, or at many other points in the fossil record.

Punctuated equilibrium attempts to explain the fossil record but fails to explain what we know about genetic mutations. Traditional neo-Darwinism restricts itself to small genetic mutations and a more plodding evolutionary pace but clashes with the fossil record. Evolutionists, in essence, face a pick-your-poison dilemma.

And beyond this, the punctuated equilibrium model itself faces a pick-your-poison dilemma. Either the proposed evolutionary bursts are too fast to be mathematically plausible, or they are too slow to explain the fossil record's pattern of abrupt appearance and stasis, even taking into account the reality that the fossil record is highly incomplete.

"And so, in the end," Meyer writes, "punctuated equilibrium highlighted rather than resolved a profound dilemma for evolutionary theory: Neo-Darwinism allegedly has a mechanism capable of producing new genetic traits, but it appears to produce them too slowly to account for the abrupt appearance of new form in the fossil record; punctuated equilibrium attempts to address the pattern in the fossil record, but fails to provide a mechanism that can produce new traits whether abruptly or otherwise."

And as Meyer further notes, it isn't just proponents of intelligent design who have concluded this: "Leading Cambrian paleontologists such as James Valentine and Douglas Erwin concluded in 1987 that 'neither of the contending theories of evolutionary change at the species level, phyletic gradualism or punctuated equilibrium, seem applicable to the origin of new body plans.'"[13]

Twenty-six years of additional research have left Valentine and Erwin still holding to this view. Here is how they put it in their 2013 book on the Cambrian explosion:

> One important concern has been whether the microevolutionary patterns commonly studied in modern organisms by evolution-

ary biologists are sufficient to understand and explain the events of the Cambrian or whether evolutionary theory needs to be expanded to include a more diverse set of macroevolutionary processes. We strongly hold to the latter position.

The patterns of disparity observed during the Cambrian pose two unresolved questions. First, what evolutionary process produced the gaps between the morphologies of major clades? Second, why have the morphological boundaries of these body plans remained relatively stable over the past half a billion years?[14]

There remains the solution proffered by Louis Agassiz a century and a half ago, that "it must have required an intelligent mind to establish them."[15] But the solution of that great naturalist did not fit with philosophical naturalism. Insufficiently materialistic in his methodology, he was pushed aside despite his superior expertise; and Charles Darwin won out despite the enormous problems that the fossil record posed, and still poses, for his theory. I am now convinced that Darwin's theory won out primarily because it fills a need: Scientism, with its allegiance to philosophical materialism, needs mindless evolution to be true, so the proponents of scientism continue to prop up mindless evolution no matter how many contrary fossils slam against it.

3. Students Begin Listening

OUR APARTMENT IN THE FAMILY QUARTERS OF HELSINKI UNIVERsity of Technology's student village was packed with students, more than sixty of them. They were sitting on chairs. They were sitting on the floor. They were sitting on bookshelves. They were sitting on tables and under tables. The year was 1976 and I had promised to talk about my doubts concerning Darwin's theory of evolution. Rarely have I had such an attentive audience or fielded so many excellent questions.

Following this, and at the request of some students, I wrote and distributed a pamphlet entitled *Evolution: Religion of Chance*.[1] If I were writing it today I would use a somewhat different style and include additional scientific findings uncovered in the intervening years, but the content is still valid four decades later. In it I cite evolutionist Julian Huxley who said that "mutations are the raw material of evolution,"[2] but also Maurice Caullery, who conceded that "it does not seem that the central problems of evolution can be solved by mutations."[3] I also described the fruit-fly experiments of Nobel Laureate population geneticist H. J. Müller. I argued that those results suggest that genetic mutations can change a species only within narrow limits.

During that same period I was acting professor of biochemistry at my university there in Helsinki, and my course lectures really stirred things up, with students crowding the lecture room and participating in animated discussions. One reason for this was that I wove through the lectures a frank discussion of the molecular-level problems facing chemical and biological evolution. Until my lectures, most of the students had been shielded from this side of things, and the peek behind the curtain galvanized their attention. I have former students who tell me that even today, some forty years later, they still remember those lectures.

At the time, the classes generated enough interest that the students organized a debate in the student union building between me and my former teacher, an assistant professor of microbiology named Pertti Markkanen. To the disappointment of many in the audience, there were no fireworks. Markkanen had never critically examined the Darwinian claims and when confronted with the evidence I presented, found himself agreeing with my arguments at almost every point. His willingness to revise his thinking when confronted by contrary evidence I find most admirable. I was gratified to see that he became a Darwin skeptic and later was very interested in the ethical consequences of materialistic evolution.

Ferment in Zürich

IN 1981 I moved from Finland to Switzerland to lead a small research team and teach at the Swiss Federal Institute of Technology in Zürich (ETH-Zürich). There, one of my duties was teaching enzyme technology as part of a course on biotechnology.

In one respect the new situation was more challenging than in Finland. The students at ETH-Zürich were, in the main, less inclined to think critically about evolutionary theory. But some of the students did

FIGURE 3.1—Friends accompany our family to the Helsinki harbor on our way to Switzerland at the end of July 1981.

ask tough, thoughtful questions about my views, and this allowed me to sharpen and refine my analysis.

One topic covered in the enzyme technology course was industrial enzymes and their modification. At the time the first genetically modi-fied protein-degrading enzymes had just hit the market. The company responsible for the innovation was California-based Genencor, and later I was hired as the research director for Cultor Limited, a 50% owner of Genencor. I knew enzymes, and the students knew I knew enzymes, so when I lectured on the barriers to the evolutionary origin and diversifica-tion of enzymes, it carried weight.

As rumors spread among the students about this controversial aspect of the course, the class numbers grew from about fifty to seventy plus. My students discussed and debated the issue of enzyme evolution in and out of class, and some of them later joined my research group for their masters and doctoral work.

This sort of reaction is to be expected. When a professor broaches a topic that is generally treated as off limits, students inevitably perk up and lean in. Science students also are attracted to scientific controver-sies and unsolved problems. What young ambitious scientist wants to be told that she has arrived on the scene after all the excitement of debate and discovery has passed and all that's left to do is dot the i's, cross the t's, and memorize things? And yet that is how origins biology is often taught, so eager are the defenders of materialistic science to guard the citadel of evolutionary theory from unwelcome questions. It doesn't have to be this way.

Mutants Good and Bad

To UNDERSTAND why I came to doubt that genetic mutations could ac-cumulate so as to evolve fundamentally new forms in the history of life, we need to wade a little further into the science of genetic mutations. A genetic mutation is a copying error in the unbelievably fine-tuned genetic system with its complex regulatory networks and protein-protein inter-actions. Most mutations are either neutral or harmful. But there are rare cases where a mutation is helpful, at least in certain limited circumstances. How could an error be beneficial, and to what degree could such beneficial mutations accumulate and lead to more dramatic evolutionary changes? To get at those answers let's look at some common mutation mechanisms and two beneficial mutations often cited as evidence for the creative power of evolution.

In 1998 I was conversing with an internationally known geneticist who specialized in finding specific sickness-causing genes in humans. According to her, science had by that point uncovered practically everything there was to know about genetics—only details remained to be filled in. It's difficult to exaggerate how wrong that geneticist was. Our understanding of genomes, genes, and embryonic development has changed dramatically during the last fifteen years, and the rate of discovery has exploded.

When I translated *Evolution: A Critical Textbook* from German to Finnish in 2000, there was a lot of discussion as to how many genes humans have. The mainstream view was that there were probably about 100,000. Some years later, as part of the human genome project, it was discovered that we only have roughly 20,000 genes. Then it was discovered that our genes are much more complex than earlier thought. A gene can contain many messages which are read in both directions; genes can overlap; genes can be split in pieces that can be joined together in different ways; and the message can vary depending on where the reading of the gene starts. Theoretically a gene can help produce thousands of proteins and numerous regulatory elements. What the geneticist colleague thought was a field transitioning out of its discovery period had in reality merely taken its first steps into a wild west of scientific discovery, one we are still in today.

We have much still to discover. At the same time, the information-rich nature of the genome is coming into better and better focus. And some of the old basics, with appropriate qualifications, still hold. In a nutshell, DNA is written in a four-character alphabet along the spine of the DNA's double-helix molecule. DNA's four-character alphabet codes for the twenty-amino-acid alphabet used to build all kinds of different proteins. A gene is a relatively short stretch of DNA that serves as a basic unit of heredity. We can think of genes as biochemical sentences whose messages are read and converted into functional proteins—the paragraphs if you will.

Proteins make up much of our body. Hair, nails, muscles, and skin are formed from different kinds of proteins. Hemoglobin is a protein that transfers oxygen. Insulin is a small protein (hormone) that controls the sugar level in blood. Some powerful toxins like botulin are proteins. And enzymes that catalyze the biochemical reactions in our body are proteins. Each of these proteins is quite distinct from other proteins.

In my research I have studied in some depth the enzymes and biochemical mechanisms that rot-fungi use when they degrade wood material composed mainly of cellulose, hemicellulose, and lignin. A birch tree, to take one example of this process, contains about 30% xylan fiber, which is degraded by an enzyme called xylanase. (See Figure 3.2 below and Figure 10.7 in Chapter 10.)

All of these different kinds of proteins are built up out of biological information, and that information can be altered through several kinds of copying errors (mutations). So, for example, a point mutation means that one DNA "letter" (a nucleotide) is changed into another letter. This change can lead to no known effect, a modest effect, or a dramatic effect. What is known as a silent point mutation may have no effect. A stop mutation terminates the reading of a gene. A reading frame mutation affects several amino acids.

ATG GTC TCC TTC ACC... ACC GTC AGC

M V   S   F   T... T   V   S

FIGURE 3.2—Part of a xylanase coding gene (upper row) of the *Trichoderma* fungus. It contains 669 genetic letters and generates a xylanase enzyme that contains (lower row) 223 amino acids. This is a biochemical sentence, whose information can be expressed as a function: *degrade xylanase fiber.*

In plant and animal cells, DNA is tightly packaged into thread-like structures called chromosomes, which can contain thousands of genes. A substantial stretch of a chromosome can be removed or added in chromosome mutations. In another type of mutation, the reading of a gene can change direction, causing it to be laid down backwards, much as you might flip a word to get a new ordering of letters. So for instance, flipping the word *rat* gives us *tar*, and flipping *fly* gives us the gibberish word *ylf*.

A gene also can accidentally get copied twice. A gene, in other words, can double. This and point mutations are considered particularly important for evolution.

In what is known as horizontal gene transfer, organisms (usually bacteria) can obtain a gene from another organism. Some geneticists see this as an underappreciated driver of evolutionary change.

Mutations can be caused by chemicals, radiation, or extreme heat or cold. But in nature mutations generally occur spontaneously for no clear reason. More to the point, mutations are rarely beneficial, and cells generally work to keep the number of mutations as few as possible. A Nobel Prize was given in 2015 for the discovery of this error-correcting system. It's as if every cell has its own personal copy editor. This copy editor is not perfect, but it is extraordinarily effective, and essential. Without it our fertilized egg cells would die long before developing to an embryo. Thanks to this error-correction system, only about one mutation for every ten billion DNA letters is inherited by the next generation. If we could hand-copy the more than four million letters in the complete plays of William Shakespeare with the same speed and accuracy that bacteria read and copy their genomes, we could dash off some 200 copies of all his plays in twenty minutes with only a single typo in just one of the 200 copies.

But even this much mutational change gives evolution something to work with, so the question is this: Could this trickle of random genetic mutations, filtered by natural selection, have generated all that new genetic information needed to build the many new kinds of plants, animals, and microorganisms that have emerged in the history of life?

Breeding efforts that use random mutation techniques, including powerful mutagenic agents such as radiation and chemicals, have long been pursued to increase the productivity of plants and animals. Those efforts have not been very successful, but there are examples of mutations that do cause us to sit up and take notice. One grows in my own garden—a cultivar of silver birch valued for its decorative wood. I was promised five dollars per kilogram if I cut it! Other examples are color variants of flowers, albino forms, blind cave fish, and of course the large number of genetic diseases in man.

Let's look now at two commonly cited examples of evolution by genetic mutation, ones that can be witnessed in the lab, and see how far such mutations can take us.

Antibiotic Resistance

SOME YEARS back I was discussing evolution with a famous geneticist in Finland. She considered antibiotic resistance as the best evidence for evolution. Antibiotics are chemical compounds mainly produced by fungi or bacteria. They interfere with bacterial growth without directly harming humans. Antibiotics may stop the synthesis of cell walls, protein or DNA. We use antibiotics, of course, to fight bacteria harmful to humans and livestock. Unfortunately, but also intriguingly, many bacteria that threaten our health have grown resistant to several if not all known antibiotics.

If we have a bacterial population with no antibiotic resistance and add an antibiotic, our experience is that one or more mutations will eventually occur that confer antibiotic resistance. For instance, an enzyme called β-lactamase makes penicillin inactive. (See Figure 3.3.) There are many modified versions of penicillin. These include G-penicillin, methicillin,

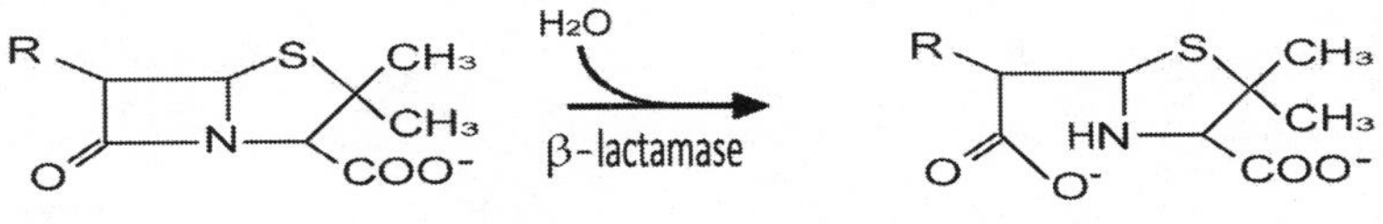

FIGURE 3.3—β-lactamase produced by an antibiotic resistant bacterium degrades penicillin (on the left) to penicilloic acid.

ampicillin, and amoxicillin. But often one point mutation in the lacta-mase enzyme is enough to make a bacterium resistant to one of these modified forms of the penicillin antibiotic.

These antibiotic resistant bacteria may overcome the antibiotics in either of several ways. They may destroy or modify the antibiotic so that it's no longer functional. They may modify the bacteria's antibiotic receptor so it does not even receive the antibiotic. They may prevent the antibiotic from entering the cell. Or they may pump the antibiotic out of the cell.

Two basic types of bacterial mutation are: (a) mutation to an existing gene; (b) the transfer of a gene from one organism to another (which is known as co-option). Both types come into play in the development of antibiotic resistance. Sometimes a mutation to an existing gene in a bacterial cell changes either the antibiotic receptor or the enzyme structure so that it is effective against an antibiotic. And sometimes the antibiotic-degrading information is borrowed from another organism. Both cases are noteworthy. But notice that neither case forms a new biological structure. They don't even create a new kind of protein. They simply tweak or transfer information that is already there.

However, it has become evident that antibiotic resistance is not something that is only a result of recent overuse of antibiotics. Low levels of antibiotics have always been around in nature and low levels of resistant bacteria have also been around. Presumably the bacteria mutated long ago to develop antibiotic resistance.

The habitual presence of antibiotics selects for those resistant bacteria until they dominate the scene and pose a clinical problem. Cognizant of this, regulators in Finland and many European countries have banned the practice of adding antibiotics to animal feed. Without the near-constant presence of antibiotics, resistant bacteria tend to fade back to their minority status in a bacterial population. This occurs because in the absence of an antibiotic threat, resistant bacteria go from superstars to bench-warmers. Due to something compromised in their genomes, they aren't as efficient or otherwise as competitive as the regular bacteria when the antibiotic isn't around, so in those cases their proportion in the bacterial population remains low. In essence, they have a specialized skill at the cost of overall fitness. This reality suggests limits to how far they could continue to evolve.

Antibiotic-resistant bacteria, then, are fascinating from a scientific standpoint and challenging from a clinical standpoint, but they aren't all they're cracked up to be as icons for the powers of evolutionary change. Antibiotic resistance appears to occur within very strict limits.

Xylitol-Eating Mutant Bacteria

ANOTHER INSTANCE held up as powerful evidence for the creative power of mutations involves the bacterium *Aerobacter aerogenes*. Normally it cannot use for energy a five-carbon sugar alcohol known as xylitol. If the bacteria are transferred to a solution where xylitol is the only energy source, the colony dies. However, once in a while some *A. aerogenes* cells survive and start to "eat" xylitol. How is this possible? Is this caused by a macroevolutionary change? This is what a colleague from New Zealand asserted when we discussed the topic in the mid-1980s (see Chapter 9). When this phenomenon was discovered in 1964, it was not clear whether a new enzyme activity had evolved, but studies in subsequent years revealed the following: *A. aerogenes* can use D-arabitol and ribitol as carbon sources. They both closely resemble xylitol. (See Figure 3.4.) In the presence of ribitol the bacteria synthesize an enzyme called ribitol dehydrogenase that can oxidize both ribitol and xylitol. However, the bacteria do not recognize xylitol, so the enzyme production is not induced and the bacteria die. However, when a mutation occurs in the regulatory region of the enzyme-producing gene, the bacteria start to produce the enzyme continuously. Such a mutant can now grow and multiply in a xylitol-containing nutrient solution.[4]

This mutation can be considered positive for the bacteria in such an environment, but genetically it is a mistake in the control system of the bacteria. The induction mechanism has been destroyed. This means that the mutated bacteria now have no control to limit the overproduc-

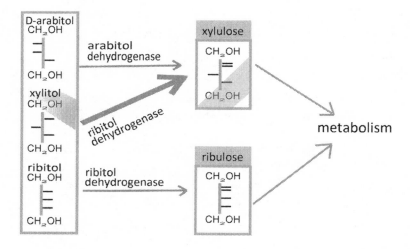

FIGURE 3.4—*Aerobacter aerogenes* bacterium can use both D-arabitol and ribitol because these sugar alcohols induce the production of an oxidative enzyme. The bacterium does not recognize xylitol. However, ribitol dehydrogenase can oxidize xylitol to xylulose. A mistake in the control region of the gene results in continuous production of the enzyme enabling the use of xylitol.

tion of one of their enzymes. Enzyme production costs energy, so this type of mutant bacterium would be less fit in normal conditions since it uses some of its energy in producing something it does not necessarily need. The mutation degraded part of the bacterium—broke it—but in so doing it conferred a niche function that the bacterium didn't before possess—a function useful under the right circumstances. However, no new genetic information was created. For random mutations and natural selection to evolve all of the many life forms we find on our planet, mutations must do more than degrade existing biological function in ways that confer niche advantages. Instead, they have to build fundamentally new forms and functions. That isn't what we're seeing in the case of *A. aerogenes*. It also isn't what we see in the case of mutations that confer antibiotic resistance.

These two examples of spontaneous mutations show that bacteria have survival mechanisms for abnormal conditions. The basic genetic system of bacteria allows them to develop new degradation and synthesis routes that improve their chances of survival. This kind of adaptation is based on existing metabolic pathways and control systems. It has nothing to do with macroevolutionary changes. If evolutionists wish to offer evidence that something more than microevolutionary change is possible through blind material processes, they will need to look elsewhere.

4. Professors and Presidents React

Internationally renowned Finnish pianist-composer Ralf Gothóni (see Figure 4.1) called me in spring 2003 and asked if I would be willing to come to talk about biological information at the chamber music evening of the famous annual *Savonlinna Opera Festival*. I had spoken at the chamber music evening in previous years, and the experience had been very pleasant, so I happily accepted the invitation. Gothóni had already invited a science popularizer and astronomy professor named Esko Valtaoja to come. He had accepted the invitation and said he had plenty of time for the event since he had vacation time. A few days later I received another call from Gothóni. He said Valtaoja had contacted him to say that he could not come because he would be on vacation!

I suspect that Valtaoja refused because he learned I would be sharing the stage with him. Earlier I had responded to an interview with him in

FIGURE 4.1—I met Ralf Gothóni on a boat trip to Germany in Autumn 2012.

a Finnish chemistry journal where he claimed that "life is nothing else than physics and chemistry—mere electricity. There is no reason to assume anything supernatural."[1] In my written response in a subsequent issue of the journal, I made every effort to be civil, but challenging his philosophical materialism in a science journal apparently had put me beyond the pale.

Valtaoja having backed out of the chamber music evening event, Gothóni then tried to invite another Finnish professor, Anto Leikola, but he also refused when he heard that I would be present. He explained that doctors had advised him not to get into situations where he could get too excited. Gothóni was left wondering why these professors were afraid of me.

Such reactions do not surprise me anymore. I first realized how sensitive the topic is when an editor asked me to write a lead article for a popular magazine. The magazine issue would focus on evolution, its philosophical basis and conclusions, with several authors contributing. The magazine *Ajankohtainen* printed my article in a 1979 issue. Altogether there were 30,000 copies, which is a lot for a nation with half the population of the state of Michigan. My article was titled "Riddle of the Origin of Life."[2] In the journal, five Finnish scientists, some of them my own professors from my days as a graduate student, evaluated my text very positively.

But that was only the beginning. The magazine issue aroused so much attention that one afternoon a radio reporter was sitting in our living room to interview three of us who had contributed. The interview was part of a June 2, 1980 science program for the Finnish public service broadcasting company Yle. Two other scientists were in studio to comment on our recorded interview, and one of them attacked us with such fury that when I heard it from my armchair at home I could hardly breathe. My wife wondered what was happening. The professor had taken offense at my claim that we scientists have no idea how life could have emerged in the first place through an unguided natural process. And in his fury he said that the basic law of nature, the second law of thermodynamics, clearly says that order increases in nature! The claim runs so counter to what the law actually says that I was stunned. I would have loved to respond to the absurd claim, but the format of the program didn't allow for it. My comments had been recorded earlier, and I had not been invited to participate in the studio discussion.

Later I sent a copy of the journal to a well-known chemist, Professor Birger Wiik, who was part of the international team studying the composition of the first Moon samples. He invited me to his laboratory, and for a moment I was able to hold moon rock in my hand. He had read my article and completely agreed with my assertion in the article that we have no idea of the origin of life. Later I got a telephone call: I heard the voice of an old man who said he was Professor Sven Segerstråhle. I knew the name. He was a noted biologist. He called to say that he had read my article and to encourage me to swim against the stream. He also agreed with me: Science knows nothing of the origin of life and the rest of evolution. I was surprised because I had studied in high school the biology textbook he had written with another Finnish professor. Evolution was extensively discussed in the book.

Another scientist who came to my defense was Professor Jouko Virkkunen. He had taught me physics and electro-engineering at Helsinki University of Technology. I discussed evolution with him, and he agreed to submit a comment on my article to a magazine. He was a deep thinker and understood very well the difficulties with evolution. He deftly summarized the reason for his skepticism toward Darwinism. "I can understand that a creature without hands can obtain a kind of lump on the side as a result of mutations," he said to me, "but I do not understand how a random mechanism can produce fine mechanics, a control system, and the computer program in the brain to move the hand. Evolution stands on clay feet." He repeated the point in front of television cameras in spring 1981.

Professor Matti Nuorteva, from the University of Helsinki, was another source of encouragement. He is known for his insect studies. He also visited my home and encouraged me as an experienced scientist to continue asking good questions.

So, there were the scientists who disagreed with me and ran from a public conversation about the ideas. There were the scientists who disagreed and practically frothed at the mouth as they denounced me. And there were the various scientists who encouraged me, came to my defense, and created opportunities for me to speak. But there was a fourth category of scientist, most rare: the scientist who disagreed with me but had the courage and equanimity to discuss our differences in a civil manner, publicly and privately.

Nuclear physicist Professor Kalervo Laurikainen was one such man. He had aroused much attention for his writings on the relationship between science and faith. One evening Laurikainen visited our home together with A. E. Wilder-Smith and his wife, and the whole evening we discussed biological information. Laurikainen had no explanation for the origin of information but could not accept the idea that biology with its information content directly points to an information source, a designer. According to Laurikainen, modern physics with its uncertainty principle leaves space for God but gives no direct evidence of design. Since we had critiqued his view publicly, he wrote an article in *Kristityn Vastuu*, a Christian newspaper, strongly favoring evolution.[3] However, he avoided the vitriol and name-calling that was such a popular tactic among many of my critics. His type, alas, is all too unusual today.

The Eclipse of Academic Freedom

THE ATMOSPHERE in our universities is now completely different from that of the open discussions that were common in the '70s and '80s. Today naturalism controls the universities so completely that debates about the problems of evolution are rarely tolerated. A good example was the National Science Days in 2009 in Helsinki University. The theme was evolution and the days commemorated Darwin's anniversary. No critical comments about the theory were allowed. Although some people suggested to the organizers that I should be invited, the suggestion was rejected.

That is passive suppression of debate, an all-too-common tactic. Then there are the calls for a more active and aggressive approach. In the journal *Acatiimi*[4] two professors went after me and a philosopher critical of modern evolutionary theory, professor Tapio Puolimatka. In the article the authors, professors Esko Länsimies and Markku Myllykangas, said that the two of us should be blacklisted and thrown out of our universities. They suggested that this should happen as soon as the science establishment gets serious about purging superstition from its ranks. In articles published in some newspapers around Finland, they urged similar tactics.

Less Than Presidential

IN THE summer of 1980 I was sitting in the office of University of Helsinki President Nils Oker-Blom with organic chemist A. E. Wilder-Smith, who was spending his summer vacation in Finland. The meeting was warm and Wilder-Smith and Oker-Blom chatted for a long time. Wilder-Smith told about his *Pro Universitate* lectures in various European universities, and Oker-Blom asked whether Wilder-Smith would consider also lecturing in Finland. He agreed to do so and the President asked me to contact one of the physics professors, Kalervo Laurikainen, who was planning an interdisciplinary lecture series for the following spring.

Spring arrived and Wilder-Smith returned to Finland directly from a lecture tour in the United States. He lectured at several universities here in Finland on evolution, the origin of life, and drug abuse. Almost without exception the lecture rooms were full and the lectures well-received. The lone exception was the Viikki campus at the University of Helsinki. Professor Matti Nuorteva invited Wilder-Smith to give a lecture as part of a seminar series to the faculty of agriculture and forestry. In the middle of Wilder-Smith's lecture, one of the leaders in the academic socialist society started to shout and rage that this was a NATO conspiracy. The interference lasted for several minutes. As a perfect English gentleman, Wilder-Smith sat down and listened to this outburst politely till the audience intervened and asked the heckler to be quiet. People had come to listen to Wilder-Smith, not to him.

Many newspapers wrote about the visit, some featuring the story on the front page. National television wanted to interview him if I agreed to participate in a live TV discussion. I agreed and met evolutionist professor Anto Leikola for the first time. At the beginning of the program the interviews of the previously mentioned professor Jouko Virkkunen and of Wilder-Smith were shown and then Leikola and I discussed them. Leikola dismissed the arguments of both professors and said the evidence for evolution was overwhelming. I pointed out the huge information problem evolutionary theory faces, but he paid no attention to it and kept repeating that the issue is settled. At that time, he was an experienced TV figure and I was a young man making my first television appearance, so there was little hope of my making a dent in his wall of nothing-to-see-here denials.

In the aftermath of Wilder-Smith's visit, President Oker-Blom, to my surprise, said he had nothing to do with Wilder-Smith's invitation. My best guess is that he was intimidated by the firestorm surrounding Wilder-Smith's lectures and was not ready to take responsibility for having invited him. That was frustrating, but I still counted the event a success. Wilder-Smith's lectures generated a tremendous amount of discussion, with a follow-up study finding that altogether 407 letters to the editor concerning the events were sent to various newspapers, of which 203 were published.[5]

I was later invited to write an article with the title, "The Worldview Aspect of the Theory of Evolution." It eventually appeared in the yearbook of the society of Christian medical doctors. Here is a short section from the piece:

> I think that one of the key factors in advancing science is to make brave hypotheses. In this respect I consider the evolutionary hypothesis fruitful. It makes some basic assumptions about the nature

of the universe. To give a hypothesis the certainty of a natural law and proclaim it as a certain result of natural science is not, however, acceptable. Exactly that has happened to the theory of evolution. Natural science should study and explain the mechanisms of nature, not establish absolute truths.

Those for whom evolution is only a scientific hypothesis can calmly discuss its weaknesses. Fortunately, most scientists belong to this group. However, those for whom evolution forms part of their worldview react almost without exception strongly and emotionally against anything that might render its role questionable.[6]

Lennart Saari, a noted ornithologist from Helsinki, who has published more than a hundred papers on birds, can testify to that latter point. In 1979 and 1980 he debated Dr. Anssi Saura, an emeritus professor of molecular biology from Umeå University in Sweden. The two faced off first at Helsinki University and then in a television debate. Saura started the debate at the university by comparing Saari to the Libyan Muslim dictator Muammar Gaddafi, presumably because both believed in God or something. Sadly, the theologians in the audience applauded for Saura.

It was not the last time Saari would be attacked for his doubts about Darwinism. He jokingly explains that due to his critical attitude against evolution, his career path "started to go upwards until it was completely vertical." In other words, thanks to his refusal to support Darwinism, his career promotion path had become so steep that it was impossible to advance further.

He soldiered on, however, and we found ourselves partnering in another effort several years later. In 1998 Timo Linnakylä, a project manager for the University of Helsinki's Palmenia Center for Continuing Education, contacted me to organize a seminar about evolution. I promised to serve as a chairman for the meeting and give a talk on the origin of life. We also invited Saari as well as Siegfried Scherer, an internationally respected scientist in the field of microbial ecology and a professor at the Technical University of Munich.

FIGURE 4.2—Siegfried Scherer, a German biologist and professor of microbial ecology at the Technical University of Munich, Weihenstephan, where he also served as the managing director of the Nutrition and Food Research Center.

One day before the seminar I got a message from the president of the university, professor Kari Raivio. He had been asking about the seminar and announced that it would be cancelled if an evolutionist was not given the opportunity to present an opposing view. Linnakylä described the meeting with the president that took place the morning before the seminar:

> Kari Raivio had sent an email to the head of the Centre. An emergency meeting was called together for the morning before the seminar. My immediate boss was present. He called me early morning from the meeting and told me that I'd better invite a supporter of evolution as a speaker or I might lose my job. I do not know if he was serious. After some telephone calls I managed to get a professor of philosophy as an additional speaker.

Disaster averted. As for the event itself, I was surprised how well-attended the seminar was despite a considerable entrance fee. More than 150 came to listen and learn, and the Palmenia Center had rarely if ever gotten such positive feedback from one of its education events. 84% of the audience rated the seminar speakers as excellent and 96% rated the content as excellent or very good. "I especially liked professor Scherer's approach to treating his material scientifically and in an unbiased way, limiting his talk to things that can be verified and avoiding speculations of things that cannot be experimentally verified," wrote one participant. As for Lennart Saari, he proved his expertise when Scherer was discussing the variability of birds. I could not interpret from German the names of the birds, so Saari helped me. For a moment it seemed that he knew every bird in Finland by first and family name!

Notwithstanding the unusually enthusiastic response to the event, Linnakylä's boss called him in afterwards and, according to Linnakylä, insisted that if this sort of seminar were organized in the future, there would need to be speakers present who represent the "commonly accepted scientific view."

Fine. In 2003 I suggested we organize a subsequent event that did precisely this. We planned together a two-day seminar entitled "Evolu-

FIGURE 4.3—The "birdman" Lennart Saari in the snow.

tion, Intelligent Design, and the Future of Biology." I served as the event's chair, and for speakers we invited two leading Finnish evolutionists, professors Anto Leikola and Petter Portin. To present the design view, we invited Drs. Richard Sternberg and Paul Nelson from the United States. Brochures were printed and invitations sent out to biology teachers. Then some professors demanded that the university's new president, Ilkka Niiniluoto, cancel the seminar. The pressure was so great that the head of Palmenia cancelled the seminar at the request of Niiniluoto. In a letter to Nelson and Sternberg, the president explained the cancellation by saying that this type of seminar would have been better suited to the university's philosophy department.

The reality was plain, however. Organizing a university event that presented both sides of the evolution/design controversy was no longer good enough. The goalposts had moved. Now any such seminar must present only the anti-design, evolutionary view. A debate between evolution and intelligent design was beyond the pale. Never mind about balance. What a sad, strange, constrictive stance for an institution of higher learning in western Europe! Universities are supposed to be places where all kinds of things, even controversial topics, can be openly discussed.

Before the cancellation was announced, the possibility became a topic of heated debate on the university's professor email list, with Linnakylä getting messages questioning the validity of the seminar. He argued that the theory of intelligent design has sparked a lot of discussion around the world and among scientists, and is one that intrigues many students, so it was important to explore the issue on campus. He went so far as to insist that it was the university's duty to introduce students to new ideas and perspectives, including ID, and suggested that such an approach would help students understand controversial issues more deeply and foster critical thinking skills.

Linnakylä's comment led to an exchange of more than 300 email messages. Some insisted the topic had no scientific value and demanded the seminar be canceled because "Leisola is a known creationist." One such critic characterized the organization of the seminar as an unbearable situation. Others in the exchange considered the organization of the seminar a brave and welcome move from Palmenia, and argued that to cancel it under lobbying pressure would be against the academic freedom to think. When had the university become an institution that tries to prevent people from thinking, some of them wanted to know.

Following this stormy discussion, President Niiniluoto "asked" the director of Palmenia to cancel the meeting. After that I got an email from Linnakylä: "Cancellation of the seminar is a shame to the university. The president's decision is against the will of the dean of the faculty of biological sciences."

I was in a difficult situation after this cancellation because Sternberg and Nelson had already booked their flights and many students and teachers had already registered for the meeting. I explained the situation to my boss, the president of Helsinki University of Technology (TKK), and asked his permission to organize the seminar in a reduced form at our university. He had nothing against it. That was heartening, though even this was far from ideal. Professors do not usually ask permission for seminars at TKK, since the university's robust spirit of academic freedom normally obviates the need to ask permission for such activities.

The situation became even stranger when a student started to collect names against this reborn version of the seminar. A petition signed by a little under 200 people was given to the president. As an upright man he saw no reason to cancel the seminar, which was organized in one of the major lecture halls of the university under the title "Biology: Tackling Ultimate Complexity." Nelson and Sternberg each had two lectures. About 200 people came on short notice.

The feedback from the participants was extremely positive. "Our teachers should learn from these guys," reported one from the audience. "The lectures were even more objective than I expected, just like they should be in our university," said another participant. I have gathered feedback from students for over thirty years, and the feedback from this meeting was some of the most enthusiastic I had ever seen. However, there was more feedback to come. A student magazine wrote about the seminar with the sarcastic title "God or Extraterrestrials Behind Evolution." The city's main newspaper, *Helsingin Sanomat* (*HS*), reported concerning the seminar, "At Helsinki University of Technology, God is included in the natural sciences."

I replied to this article by noting that God was never mentioned in the seminar. When my reply was published in *HS*, one of the professors in my department started a discussion on our university's professor email list. There he wrote that he considered the seminar and my letter to the editor as a personal insult, and worried that together they put the university's natural sciences division into a very strange and embarrassing light. The seminar, it was argued, was a sad mixture of religion and religiously colored pseudoscience; what a grubby little nest the university president and I had made. Various professors joined the choir. I contacted each of them to ask how I had offended them. In these one-on-one discussions the tone grew more civil and I had some very interesting discussions with my colleagues. Some even admitted that the end result was good and constructive. Some wondered at the fanatical lynching mentality of some of the emails. Others told me that they actually considered it honorable that over the preceding several years I had taken time to critically evaluate evolution and recommend intelligent design as a possible research hypothesis in biology. While many of them remained loyal to evolution, they conceded that my work critically analyzing evolutionary theory did not make me, as one of my critics in the email exchange put it, a destructive and indiscriminate defender of the occult.

This is how one senior person who attended the seminar commented about it:

> *HS* reported showily on the seminar.... I did not recognize the seminar in the article the editor wrote. The title "At Helsinki University of Technology God is included in natural sciences" is simply not truthful. Such a thing was not even mentioned. But most of all I was shocked... by the amazing prejudice of prestigious Finnish scientists and opinion leaders and their fear of being contaminated

by false doctrines. The seminar was blamed for being anti-science already beforehand without any arguments.

Nelson's and Sternberg's excellent PowerPoint presentations were on my lab's webpage, but I was pressured to remove them, which was unheard of. Also, none of those critical of the seminar was interested in the scientific content of the meeting or in the very positive feedback from the audience. None of the critics came to the seminar or requested a recording of the presentations afterwards. The power of prejudice can be overwhelming and override objective judgment.

Not long after the turmoil described above, I accepted an invitation from the medical students at the University of Turku to talk about the origin of life. I accepted, but to my surprise, the president forbade my presence in any of the university facilities. I emailed him to ask the reason for this, but he never responded. The meeting, however, took place at the home of a local student minister where almost a hundred students jammed in. I later heard that Professor Petter Portin, who had been scheduled to speak at the previously mentioned seminar, had demanded that I be disinvited.

Nine years later, I spoke to the students at the University of Vaasa (in March 2013). I planned to talk about ID, and my talk was made known to university personnel via email. Then some professors demanded that my talk not touch on the theory of evolution.

By that point I was not surprised by this sort of behavior. I long ago had come to see that those bent on intimidation think nothing of shutting down debates and marginalizing scientists while paying lip service to the value of academic freedom. But I take encouragement from the fact that such people cannot make the evidence against modern evolutionary theory disappear altogether. At most they can sweep it under the rug and hope no one is curious enough to pull back the rug. Unfortunately for them, even some scientists who do not count themselves as proponents of intelligent design have begun to do just that.

Micro vs. Macro

ALL THE micro-evolutionary changes described earlier in this book are intra-species modifications that happen within narrow limits. These changes can be seen all around us—in laboratories, farms, forests, lakes, and my own garden. Modern evolutionary theory holds that these small changes—variations within species—can accumulate and, over long ages, give rise to fundamentally new organisms. As the story goes, single-celled organisms evolved into multi-celled organisms and eventually into plants and animals. Dinosaurs became birds. And mammals living on dry land morphed into whales.

On this view, macroevolution is simply a result of long-lasting microevolutionary changes. Some go so far as to insist that the two concepts thus have no qualitative differences, separated only by time. Some even accuse evolution skeptics of inventing the micro-macro distinction out of whole cloth. But in fact, the term "macroevolution" was used by leading evolutionist George Gaylord Simpson in 1944, and before him, another important evolutionist, Theodosius Dobzhansky, used the term "microevolution" when talking about small changes within a species and "macroevolution" for the evolution of new species.[7]

So the invention of the two terms was not some devious plot cooked up by early design proponents. More fundamentally, the problem for evolutionists isn't merely a semantic one, as if the two terms were the problem and if we could just dispense with them, all would be well in the land of Darwin. No, the problem is that macroevolution is a philosophical concept starved of observational evidence.

Now, some defenders of macroevolution take umbrage at the idea that there is little evidence for macroevolution. They point to all the examples of microscopic and anatomical similarities across species, families, classes, and phyla. A striking instance of this is the fact that DNA exists in all living organisms. To many evolutionists, such common features are abundant proof of unguided macroevolution.

But in the realm of human technology we see similarities all the time among disparate technologies. Here the explanation is, of course, intelligent design. Designers pick and choose ideas and mechanisms well-suited for a particular goal. In one case the wheel is used and adapted for a water mill; in another case for a bicycle; in another, for an automobile. So, what about in the realm of living things? Might not a designer have used and reused a good design concept in widely different biological contexts?

The only way to jump from biological similarities to unguided macroevolution, then, is to rule out the design hypothesis from the outset. But if the debate is evolution vs. intelligent design, then ruling out design from the outset is just so much question-begging. It's just one more way to shut down debate and protect modern evolutionary theory from competition and critique.

This is no way to advance knowledge. Science should be about evidence, not rigged games. What does the evidence suggest is the better explanation for the origin of fundamentally new species and body plans in the history of life, blind evolution or intelligent design? And what findings might count in favor of one hypothesis or the other? Those are the kind of questions an unfettered, truth-seeking scientific culture is happily willing to tackle.

And happily, at least some leading scientists have been willing to put the theory of evolution in "empirical harm's way," to borrow a phrase from philosopher of science Del Ratzsch.[8] In 1965 one of the most important scientists of the last century, Linus Pauling, and biologist Emil Zuckerkandl, considered by some as the father of molecular biology, suggested a way that macroevolution could be tested and proved: If the comparison of anatomical and DNA sequences led to the same family tree of organisms, this would be strong evidence for macroevolution.[9] According to them, only evolution would explain the convergence of these two independent chains of evidence. By implication, the opposite finding would count against macroevolution.

So what were the results? Over the past twenty-eight years, experimental evidence has revealed that family trees based on anatomical features contradict family trees based on molecular similarities, and at many points. They do not converge. Just as troubling for the idea of macroevolution, family trees based on different molecules yield conflicting and contradictory family trees. As a 2012 paper published in *Biological Reviews of the Cambridge Philosophical Society* reported, "Incongruence between phylogenies derived from morphological *versus* molecular analyses, and between trees based on different subsets of molecular sequences has become pervasive as datasets have expanded rapidly in both characters and species" (emphasis in original).[10]

Another paper, published the following year in the journal *Nature*, highlighted the extent of the problem.[11] The authors compared 1,070 genes in twenty different yeasts and got 1,070 different trees. An article in *Quanta* magazine, reporting on the paper in *Nature*, highlighted the challenge these findings pose for the Darwinian tree of life:

> According to a new study partly focused on yeast, the conflicting picture from individual genes is even broader than scientists suspected. "They report that every single one of the 1,070 genes conflicts somewhat," said, Michael Donoghue, an evolutionary biologist at Yale who was not involved in the study. "We are trying to figure out the phylogenetic relationships of 1.8 million species and can't even sort out 20 [types of] yeast," he said.[12]

These results aren't what we should expect from a process of blind, gradual macroevolution. The contradictions vanish, however, on the design hypothesis. That is, the experimental results are not out of place if the living world is the result of a designing intelligence selecting and adapting design concepts for use in a variety of design blueprints. Many scientists, of course, consider the design hypothesis off-limits, but interestingly, among these there are now some prominent figures who have abandoned neo-Darwinism's mutation-selection mechanism, having decided that it is unable to produce macroevolutionary change. One of these is the University of Chicago's James Shapiro. Here is how he expressed his skepticism:

> One of the most important questions in evolution is: How can new adaptations originate? This is a difficult question, because most evolutionary novelties, such as the eye or the wing, involve the orchestrated expression of many different loci, a number of which act in the expression of multiple phenotypes. Conventional explanations that randomly generated advantageous changes in complex characters accumulate one locus at a time are unconvincing on both functional and probabilistic grounds, because there is too much interconnectivity and too many degrees of mutational freedom.[13]

Natural Selection as a Creative Force

THE LABORATORY and fossil evidence for macroevolution by chance mutations and natural selection has proved disappointing, but this hasn't stopped the idea of natural selection from penetrating much of our culture. Nor has it stopped many evolutionists from putting great faith in the power of natural selection.

"Natural selection has given the boreal owl uneven ears so that it can better locate a mole in a dark forest at night; for man natural selection has given brains that are able to think," explains one Finnish-language textbook.[14] The confidence of the pronouncement is typical. My children's English, physics, history, and biology textbooks all tell in various ways about evolution, and often weave in discussions far removed from strictly scientific ones. The teacher who trained my kids to drive used the concept of natural selection. And a consultant friend advised me to use the concept of natural selection when I had to reduce the number of personnel while working as a research director in a large biotech company. "The best remain, the weak are selected out," he said. The idea seems to be everywhere.

Darwin's primary evidence for natural selection was plant and animal breeding. And there are many impressive cases that can be cited from this pool. Europeans began cultivating sugar beets during the reign of Napoleon. Originally sugar beets contained about 5% sugar. After intensive breeding and selection sugar beets today contain about 20% sugar. This was made possible by selecting in every generation those healthy varieties that contained the highest amounts of sugar.

Breeders have enhanced the desired properties of many different agricultural plants and animals through a similar selection process: more wool from sheep, more milk from cows, more and larger potatoes, strawberries, and grain varieties. Darwin thought that a similar process functions in nature. He compared the intelligent breeding process that humans have used to the blind natural process. He asked: If we have been able to produce such achievements in a short time, how much more can nature do in millions of years by natural selection?

Julian Huxley, grandson of Darwin's contemporary and defender Thomas Huxley, described natural selection as "inevitable" and "*the* only effective agency of evolution" (emphasis in original). He went on to say that it "converts randomness into direction, and blind chance into apparent purpose. It operates with the aid of time to produce improvements in the machinery of living, and in the process generates results of a more than astronomical improbability, which could have been achieved in no other way."[15] But skeptics countered that Darwin's examples of plant and animal breeding actually work against his argument, first because they are examples of artificial selection rather than natural selection, and even more decisively, because those examples of evolution always took place within firm limits.[16] Dogs remained dogs. Pigeons remained pigeons. Micromutations leading to macroevolution, the skeptics insisted, were products of the imagination.

More than a century and a half after Darwin made his argument from artificial selection, the evidence is still wanting. All the textbook examples of natural selection give no evidence of a far-reaching creative power. Consider the classic example of color changes in peppered moths (*Biston betularia*). The story goes like this: Due to coal-burning in parts of England, the surfaces of trees turned dark and the color of peppered moths in the region evolved from light to dark. The reason, so the story goes, was that before the coal pollution, birds would see the dark moths

FIGURE 4.4—Light and dark forms of the peppered moth on a birch trunk.

against the white birch and eat them. When the trees became dark the light moths were eaten, the dark moths escaped detection, and the color of the moth population changed from mostly light to mostly dark. Natural selection had shown its power. Evolutionists love this story, but its credibility and explanatory power are actually quite weak. First, peppered moths do not usually rest on a naked trunk but hide among the leaves. Also, many of the moths shown in textbooks are dead and fixed with needles. And finally, even if the story were true, natural selection did not create anything new. Dark and light moths already existed, and only their relative proportions within the population changed. Another textbook example of natural selection is the beak variation of Galápagos finches. Bigger, stronger beaks are favored during dry periods when seeds are tough and difficult to break. Evolution in action, students are told. But this variation is cyclical. During rainy periods the smaller-beaked finches tend to rebound. The beak variation tends to occur within strict limits. Again, natural selection has created nothing new.[17]

Is there selection in nature? Yes, and also speciation, when populations are divided or circumstances favor survival of certain gene com-

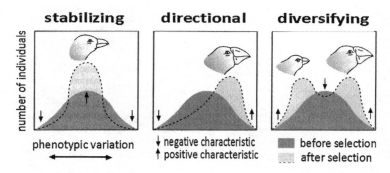

FIGURE 4.5—As a result of selection, the beak size of bird species can vary, but only within a narrow range set by the information content of the gene pool.

binations. Selection can occur in three different ways. (See Figure 4.5.) First, and most commonly, selection can stabilize the situation by eliminating individuals that might otherwise lead to changes in the population's gene pool. Second, selection can be directional, when selection pressure favors certain properties that exist but are not yet dominant in a population. In such cases the gene pool of a population tends to become less diverse. (See Figure 4.6.) Third, diversifying selection occurs in rare cases when the circumstances favor two extremes of a certain property.

But whatever the type, selection has never been known to create a significant amount of new genetic information. (At best, "back-mutations" can restore information that was lost through earlier mutations.[18]) Heritable mutations are the only way to change the genome of an individual, and as we saw above, extended research on bacteria and other rapidly breeding microscopic life have produced only modest microevolutionary variations within very strict limits. Moreover, those research results are mathematically tractable, meaning they can be extrapolated from a few decades and a few laboratory colonies to billions of years and the entire planet. The results of those mathematical extrapolations reinforce the conclusion that the mutation/selection mechanism only produces change within severe limits.[19]

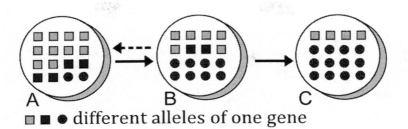

■ ■ ● different alleles of one gene

FIGURE 4.6—The figure shows how selection influences the gene pool of a population. A) Alleles of a gene (different forms of the same gene) in a certain environment X; B) the amount of any given allele changes in a new environment; C) after several generations, some of the original allele forms become rarer and may even disappear. The broken arrow shows that a return to the original pool A is possible but rare.

While lecturing years ago at the University of Helsinki, I told the audience that I do not know of a single case where the mutation/selection mechanism has created new information. I challenged the audience to help me. Maybe somebody knew of a good case. After a moment of silence, one man in the audience raised his hand and said he knew of one. His example was sickle cell anemia, a blood disease. It is caused by a single mutation in hemoglobin. The change makes blood cells crescent-shaped and causes severe anemia in those with the mutation, but the disease protects a person from the malaria parasite, which cannot reproduce in cells having this mutation. This is only a net benefit, though, when the mutation is inherited from only one parent (heterozygosis). If the gene is inherited from both parents (homozygosis), persons die young. Sickle cell anemia is not an example of the creative power of the mutation/selection mechanism. Information is not increased by this genetic mistake. It's an example of nature breaking an existing biological mechanism, a break that creates a niche advantage, a common pattern that Lehigh University biologist Michael Behe explored in a 2010 peer-reviewed essay in *The Quarterly Review of Biology*. There he notes that "the rate of appearance of an adaptive mutation that would arise from the diminishment or elimination of the activity of a protein is expected to be 100–1000 times the rate of appearance of an adaptive mutation that requires specific changes to a gene."[20]

Think of an old metal door with a flimsy lock, a door that eventually rusts shut. It's now a pretty terrible doorway, but it does have one advantage over its pre-rusted self. It is harder to break open. If you didn't need the doorway but wanted the security of that door being hard to break through, you might consider the outcome a net gain. But you would still have no reason to begin marveling about the creative powers of rust. This is akin to the sickle cell anemia mutation. Something broke and created a niche advantage. But no novel molecular machine was created, much less a novel organ or organism.

Extreme examples of selection are fish species that have lost their eyesight in underwater cave ecosystems, and insects that have lost the ability to fly while living on windy islands. Such examples are hardly cause for celebration.

Intelligent selection carried out by man leads to similar results. This can be seen in dog-breeding where genetic diseases become a real problem as the breed gets too far from the wild type (wolves). The chain can be described as follows:

Repeated selection > Gene pool becomes poorer > Variability decreases > Adaptability to environmental changes decreases > Danger of extinction increases

At first glance dogs may seem like a promising counterexample, even if compromised by the fact that they're a case of artificial selection rather than natural selection. But setting that issue aside, we are confronted by a breed such as the greyhound. It is faster than its wolf ancestor over short distances. These racing dogs running at top speed are a wonder to behold. However, greyhounds—like other types bred to excel at some particular task—are specialists. Their breeding has sacrificed overall fitness in pursuit of a niche advantage. Wolves are vastly more fit to survive in the wild than are greyhounds.

So, given the mounting evidence against the powers of natural selection, why does all the optimistic talk about natural selection persist in the scientific community and the popular media? I'm convinced it's an outgrowth of the materialistic paradigm. Those who adhere to the paradigm will not consider the possibility of intelligent design, and they understand that blind evolution is the lone alternative for explaining life's diversity. Most of them are also convinced that blind evolution requires some version of Darwin's random variation/natural selection mechanism if it is to succeed. Boiled down to its essence, the logic is simple if starved of evidence: Intelligent design must not be true, so the chance/selection mechanism must be adequate.

5. PUBLISHERS HESITATE

IN EARLY January of 2002 I was sitting in the meeting room of my laboratory with a small team of five specialists. At the urging of an editor for the biggest publishing house in Finland, WSOY, we agreed to create the first Finnish-language biotechnology textbook for polytechnic schools and first-year university students. When the project was nearing completion, my students were able to vote for the cover picture. It was an exciting time. My responsibility had been to guide the project, keep the timetable we had agreed on, and write the last chapter. It dealt with regulatory, safety, and patent issues, as well as ethical questions. Here is an excerpt from that chapter, a passage I was sure would stir the pot:

> In order to have correct ethics for biotechnology, we must have the right view of man. We can illustrate the importance of worldview by comparing two opposing views of man: Judeo-Christian and Darwinist-materialist. Well known Darwinian biologist Edward Wilson and philosopher of science Michael Ruse have said, ethics is an illusion, caused by our genes.[1] The problem of this starting point is that real discussion of ethical norms is meaningless. Because there is no final basis for ethical norms, the discussion is controlled by varying opinions of man.
>
> Contrary to Darwinist ethics the Judeo-Christian basic premise defines humans so that humans are humans from the beginning of fertilization. This means that human rights and protection belong to him from the beginning.[2]

When the textbook appeared, I was pleased to see that the textbook was adopted in polytechnics and some universities across Finland. By now, hundreds of students have studied the book as part of their training in biotechnology. To my surprise there was little reaction either for or against the passage contrasting the materialist and Judeo-Christian foundations for ethics and human rights. I was surprised because, in my experience, any published criticism of Darwinism received immediate, often shrill, and occasionally unhinged reactions from dogmatic Darwinists.

For example, four years earlier, I had met with some friends and a German professor, Siegfried Scherer, in an underground gallery of a famous Finnish artist, Kimmo Pälikkö. From this meeting grew the plan to translate into Finnish the German original text of *Evolution—Ein Kritisches Lehrbuch* (*Evolution: A Critical Textbook*).[3] The book, now in its seventh edition, is the result of nearly thirty years of work by several German scientists. I consider it one of the best critical scientific analyses of evolution. Translating the work of so many specialists turned out to be a daunting task, but with the help of many Finnish experts and after two thousand hours of labor, the text was ready for printing.

I contacted some major Finnish publishers, but none was ready to publish the book. Too complicated, we were told. Wouldn't sell much. This led to us starting our own small publishing company with a friend of mine. The book was published in the summer of 2000 and has done well enough to see a second edition.[4]

The first reviews came quickly. One Christian magazine noted approvingly that the textbook offered a good survey of "the contradictory opinions of the science world concerning the origin of the world and presents a convincing number of evidences against the traditional theory of evolution."[5] One daily newspaper reported that "Erkki Ranta, a professor of zoology from Helsinki University, considers the facts of Leisola's book, based on a quick reading, correct" and judges it "suited to a critical reader" but "criticizes its conclusion that God is the creator of everything."[6]

Finland's largest newspaper, *Helsingin Sanomat* (*HS*), grudgingly described it as a "more professional treatise of the discussion about evolution than many creation reflections" but then quickly warned that the book "reaches a completely new level of seeming like science" and that "this is only a likeness, aping science without its own content."[7] True to form, the newspaper did not even consider the book's arguments.

Their blinkered response was frustrating, but not unexpected.

I didn't respond to them on this occasion, but I did so not long after, when they betrayed a similar bias while writing about the Darwin anniversary and a list of more than 100 Ph.D. scientists expressing skepticism of neo-Darwinism.

I sent the newspaper the following letter:

> Instead of opining, the reporting on science should at least try to be based on facts. The science editor in your paper is guilty of the former when writing on Darwin's anniversary and about "doctrines that reject Darwinism." *The New York Times* reported on the same topic (April 8, 2001) but chose fair reporting, giving to both Darwinists and critics a chance to give their arguments. *The Washington Times* did the same in October 2001 when reporting on the signed letter by a hundred scientists: "We are skeptical of claims for the ability of random mutations and natural selection to account for the complexity of life. Careful examination of the evidence for Darwinian theory should be encouraged."
>
> One of the signers was Professor Henry Schaefer III, who has published some 800 scientific papers on theoretical chemistry and has been five times a candidate for a Nobel Prize. "Some defenders of Darwinism," Schaefer said, "embrace standards of evidence for evolution that as scientists they would never accept in other circumstances."[8] The matter is a scientific discussion, which has implications for worldviews—not a controversy between religion and science.

Instead of publishing my letter, the newspaper sent me an oddly worded email from the newspaper's upper management:

> Stop developing these half-factual pseudoscientific writings and tergiversations. If you do this tell directly, at the same time, that you are good believers. It would be better for us readers and Finns, and all round much easier. Use bravely the word *Jesus*. You know very well that Americans have to explain things to please believers all the time because there the readers attack. Like you.

In Britain no newspaper listens to believers with weak arguments ahead of the evidence for evolution.... the whole world heard
again after the 11th of September in the name of all kinds of heavenly dudes [sic]. Who enjoys that when both sides fight in the name
of god?

I've received angrier letters, and maybe even more obnoxious letters,
but that one may take the cake for sheer strangeness. Despite being unsure of the writer's mental clarity, I decided it was important to respond
to the writer and the paper:

In the eyes of most people, science is the neutral search for truth.
Therefore it is important that you report about science correctly. I
have never hidden my worldview when I speak about such frontiers
of science like evolution. You are guilty of mixing scientific observations and a materialistic worldview—which is its own religion—into
a soup that presents materialism as equal to truth.

While living in Zürich I listened to a lecture series organized by
science philosopher Paul Feyerabend, and I partly agree with him
concerning the arrogance of science in trying to steal the whole field
of rational thinking: "Science is much closer to myth than a scientific philosophy is prepared to admit. It is one of the many forms
of thought that have been developed by man, and not necessarily
the best. It is conspicuous, noisy, and impudent,... that most recent,
most aggressive, and most dogmatic religious institution." (Quotation from *Against Method*)

On another occasion, we wanted to advertise the ID seminar mentioned in Chapter 3 in a Finnish science journal, but received the following reply: "Our journal deals with science. We do not want to advertise a
seminar where a non-scientific worldview is promoted. Thus, our journal
will not publish your advertisement."

Notice the slip. They didn't complain about a non-scientific methodology, which is the standard line when complaining about intelligent
design. No, they claimed that ID was promoting a "non-scientific worldview." Here they were perhaps more honest than is prudent for their
cause. The approved worldview, in their eyes, is philosophical material-

ism. To their way of thinking, this alone is scientific—scientific materialism.

And because they hold to their worldview dogmatically, and have surprisingly weak arguments for their position, they tend to react defensively when challenged.

A reporter for a Finnish afternoon daily encountered this pattern when he interviewed several Finnish scientists about the origins controversy. He interviewed me as a skeptic of evolution, and several Darwinian loyalists. He titled the resulting article "Criticism of Evolution Increases: School Children Get Outdated Information." The outdated information he may have been referring to were things like the fudged embryo drawings of nineteenth-century Darwinist Ernst Haeckel, drawings that have long since been debunked and rejected even by mainstream evolutionists. But pro-evolution biology textbooks continue to recycle these and other discredited evidences of evolution.

The newspaper reporter detailed this in his article, along with a description of how some of the pro-evolution scientists responded defensively when he asked them tough questions. Unfortunately, the editor-in-chief was unwilling to publish the well-written and factual article. Here is a short excerpt from the piece:

> I have now tried for two weeks to clarify what the science world thinks of the origin of life and evolution on earth, and found out the following: 1) There are many different views about the topic and 2) many experts in the field react emotionally regarding their views. Their getting angry, transferring calls to others, and hanging up the phone have shown me that many scientists have poor evidence for their own views, and some of them feel so much uncertainty that they are not able to answer the questions of a not very knowledgeable reporter.

One of my earliest encounters with this sort of prickly defensiveness was in the early 1980s. It all begin in April 1981. I was sitting with A. E. Wilder-Smith in the office of the director of WSOY. The professor mentioned his book *Natural Sciences Know Nothing of Evolution*[9] and noted that it had already been translated into some other languages. We agreed then and there that the book would be published in Finnish with me as the translator.

The task was challenging because my family and I were preparing to move to Switzerland. But I met the deadline, the book was published, and my family and I headed off to Switzerland.

I figured the evolutionists would attack it within a week of its publication. As it turned out, it took only three days before HS published a one-page critique by my sometime nemesis, Professor Leikola. (Yes, even his name sounds like a doppelganger version of Leisola!) In summary Leikola wrote in his review titled "Sermon against Evolution" that the book was lousy, the author ignorant of his subject matter, and the translation miserable.[10] To further emphasize his scorn for the author, Leikola compared Wilder-Smith to a "Sergeant-Major," an allusion to Wilder-Smith having been a consultant for NATO forces on the question of how to address drug abuse problems.

After I arrived in Switzerland, I wrote a letter to my publisher mentioning first that our fourth baby had just been born and then enquiring how the book had sold. He answered that they had sold about 700 copies. This was a respectable start for a non-fiction book published in Finnish, so I was unprepared for the letter's next bit of news. He told me, without explanation, that he had destroyed the remaining copies of the book, and advised me to concentrate on raising my children.[11]

No, they didn't hold a village book-burning for the newly translated text, but the results were much the same. I was dumbfounded by the letter, especially when our discussion in the spring had been very friendly and upbeat. Perhaps the hullabaloo around the book had scared him?

The book did have its share of professionally powerful enemies. One was professor Ilkka Niiniluoto, a philosopher and later president of Helsinki University. He had this to say about it:

> The arguments of Wilder-Smith against evolution seem to be a new version of an age-old teleological attempt to prove God using only fashionable new concepts (probability, program, code, informa-

tion): The order in the world presupposes a designer and God, and the continuous increase in order presupposes skillfully calculated reprogramming or continuous creation. The weaknesses of this argumentation are known for those who have learned in their philosophy studies the objections presented e.g. by David Hume and Immanuel Kant.[12]

Niiniluoto also said that Wilder-Smith incorrectly thought his opponents believed that life appeared suddenly, in a single step.

Niiniluoto was mistaken on all counts. Wilder-Smith never said that his opponents believed in the spontaneous one-step origin of life. Instead, he said that they believed that life first arose by undirected natural chemical reactions.

We won't offer an evaluation of Hume or Kant here. Suffice to say that many notable philosophers reject the critique, and the contemporary design argument from information differs in significant ways from the design arguments of the eighteenth century, those that Hume in particular criticized. Philosopher and mathematician William Dembski explores these matters, and makes a case against Hume's critique in his book *Intelligent Design*, which I translated into Finnish in 2002.[13]

By implying that Hume and Kant had settled the whole matter more than 200 years ago, Niiniluoto was playing one of the evolutionists' favorite games: bluffing.

Biological Information

Niiniluoto also charged that "Wilder-Smith confuses two different types of information: Physical information (the reactions between components in the cell) and semantic information that is linked to meaning in language." We disagree with his assessment, but unlike so many of the attacks on evolution skeptics—which often amount to some combination of nitpicking, bluffing, and name-calling—this charge at least has the merit of being a specific objection, and one focused on a key aspect of the design argument, so we want to respond to it at length.

In our everyday experience, we find that intelligent agents create new information (books, song lyrics, speeches, software). And we never witness mindless forces generating new information. Laboratory experiments, computer modeling, and probability mathematics all confirm that this uniform experience likely is universally the case—information is the product of mind. Based on this combination of experience, experimentation, and mathematical analysis, we can infer that the best explanation for biological information is intelligent design. Niiniluoto attempted to cut the legs out from under this chain of observation and reasoning by insisting that biological information is too different from "semantic information" to draw any inferences from the latter to the former. He is right in saying that researchers recognize at least two different information types, but mistaken in saying biological information isn't semantic information in the sense that information theorists use that term.

The information we find in computer software, and the information we find in DNA, share important features in common, and it's these common features that Wilder-Smith focused on in making his design argument. Biological information is information written in code language and programs for function. Yes, there are differences. For one, biological information and biological information-processing are vastly more sophisticated than anything any computer engineer has ever devised. But Wilder-Smith was perfectly aware of those differences and focused on what the two varieties of information share in common.

In this, at least, he was on fairly uncontroversial grounds.

The information revolution in biology started in the 1950s with the discovery of the chemical structure of the DNA molecule. Subsequent research revealed that information is packed in this molecule in coded form and translated by RNA molecules into the various amino acid types used to build various proteins and protein machines. DNA information is functional and specific. This means that the sequence of code letters is important for the desired function, in the same way that the sequence of code letters in a software program is crucial for fulfilling the program's intended function.

The forces of nature outside of living systems are good at generating randomness, such as the rubble left behind after an earthquake. These forces also are good at generating repeating patterns, such as we see in a whirlpool spiral. But random chemical reactions cannot produce meaningful language. Nor can repeating patterns such as we see in a whirlpool. This is why no software program of any sophistication is a mere repeating algorithm, such as abababab or acegiacegiacegiacegi—much less a random string of gibberish. The software-like nature of biological information (neither a random pattern nor a highly repetitive algorithm-like pattern) explains why no theory of blind chemical evolution has been able to unravel the riddle of the origin of life.

The history of life is a history of huge information increases, beginning with the origin of the first life and then at other dramatic points, such as in the Cambrian explosion, where numerous new phyla (not just new species but fundamentally new body plans) appear during a relatively narrow window of geological time. No known chemical, physical, or biological phenomenon explains such massive increases in biological information.

In Niiniluoto's defense, information theory can be confusing, with some of the confusion stemming from an idiosyncratic definition of information employed by one of the founders of information theory, Claude Shannon. Shannon was a mathematician who near the end of the 1940s developed a way to measure the information content of a message. He said it had to do with how improbable the sequence was, given all the possible sequences it might have been.

For example, if the message is a binary string of 1s or 0s, and the string is three digits long—e.g., 001—then the odds of that particular sequence occurring by chance are 1 in 8 (2 x 2 x 2)—mildly improbable. A longer binary sequence is more improbable and so contains more information. If the sequence of 0s and 1s is eight digits long (e.g., 01001011), the odds of that particular sequence is 1 chance in 2 x 2 x 2 x 2 x 2 x 2 x 2 x 2, or 1 chance in 256—more improbable and therefore containing more information.

If each digit could be any of ten different numbers (0 through 9) then the improbabilities mount much more rapidly with each additional digit.

Binary code of computers

Two symbols in groups of eight can be interpreted as letters.

A	B	C	D	E	F
10000001	10000010	10000011	1000100	1000101	1000110

DNA's four-letter alphabet

Four symbols are read in groups of three and translated into amino acids.

Stop	Pro	Ala	Lys	Ser	Glu	Arg	Cys
TAG	CCC	GCT	AAG	TCT	GAA	CTG	TGC

FIGURE 5.1—Comparison of binary code and genetic code.

So the string 834 is 1 chance in 10 x 10 x 10—that is 1 chance in a 1000. After all, the string might have been 000 or 999 or any whole number in between—1000 possibilities in all.

We can take Shannon's system for measuring information-carrying capacity and apply it to the sequence of letters in DNA. DNA, recall, involves a four-character chemical alphabet. The letters A, T, C, and G are used as symbols for these four chemicals. The probability of a particular three-character string of DNA letters occurring (see Figure 5.1) is thus 1 chance in 4 x 4 x 4—1 chance in 64.

Shannon's information measure, however, only tells us the information-carrying capacity of a given sequence. It doesn't distinguish between meaningful and meaningless, functional and functionless. The following pair of letter chains are equally improbable and thus contain equal amounts of Shannon information.

oi maamme suomi synnyinmaa

klsanm msbmnx kjkshmoyoe n

There is an important difference here that Shannon's definition of information doesn't tease out. English-speaking readers might also miss the difference. One needs a key to understand the difference, and here the key is the Finnish language. The first line of characters is the first line of the Finnish national anthem, whereas the second line is gibberish (as meaningless in Finnish as in English). Shannon information may refer to a line of meaningless gibberish, or it may refer to a meaningful sentence, like the words of a national anthem. Or it may refer to a functional line of code, such as we find in a computer program or in a stretch of functional DNA.

DNA is least like the line of gibberish and most like a computer program. DNA stores and distributes information in a cell to manufacture proteins or control the cell systems. DNA contains specified or functional information—not just information-carrying capacity. In other words, it's Shannon information and then some, much as a software program is Shannon information and then some. That extra something, invisible to the Shannon information framework, is sequence specificity or functionality.

Evolutionary biologist George Williams describes biological information this way:

> Evolutionary biologists have failed to realize that they work with two more or less incommensurable domains: that of information and that of matter... The gene is a package of information, not an object... Maintaining this distinction between the medium and the message is absolutely indispensable to clarity of thought about evolution... In biology, when you're talking about things like genes and genotypes and gene pools, you're talking about information, not physical objective reality.[14]

We would only add that information is also objective reality. It's simply *immaterial objective reality*, which can and does manifest itself in and through the physical.

Physicist and astrobiologist Paul Davies understands cellular information thus:

> The living cell is best thought of as a supercomputer... Most of the workings of the cell are best described as information... which leaves us with a curious conundrum. How did nature fabricate the world's first digital information processor—the original living cell—from the blind chaos of blundering molecules? How did molecular hardware get to write its own software?[15]

Genetic information is much more sophisticated than what we have briefly described here, and even more sophisticated than anyone yet understands. For one, biological form needs additional information outside DNA. The genome has control elements whose importance and function we are just learning to understand. More will be said on this in a later chapter; here it suffices to say that biological information is everything that computer software information is, and then some.

We can return now to Niiniluoto's claim that "Wilder-Smith's argument against evolution seems to be a new version of an age-old teleological attempt to prove God using only fashionable new concepts (probability, program, code, information)." But we talk about probability, programs, code, and information because of specific discoveries and advances in biology and information theory. And as we have already seen, design theorists aren't the only ones insisting that DNA is akin to computer information. Microsoft founder Bill Gates has said it. Leading evolutionary biologists have conceded it. And physicists like Paul Davies have marveled over it. It's a big, interesting reality that has come into sharp focus due to the creation of fields that didn't even exist in Darwin's day—molecular genetics and computer science.

Niiniluoto also described contemporary design arguments as teleological. Teleology refers to the idea that there is purpose in the processes of nature. According to evolutionary materialism there is no purpose. Recently, a Finnish physics teacher bluntly distilled the latter view when he commented publicly that "an adjustable wrench has a purpose but life has no purpose." Yes, that assessment defies our most powerful intuitions, but set that aside for the moment. There is a more tangible problem with the claim: It leads to failed predictions.

This type of evolutionary thinking has encouraged such notions as vestigial organs and junk DNA, since blind processes often produce waste and poor solutions. The theory of intelligent design sees biology through different spectacles. Both of these paradigms can be predictive and offer explanations for various biological realities. But which one better reflects reality? During the last several years a new field in biological sciences, called systems biology, has emerged. It is concerned with "why" and "for what purpose" questions. University of Pittsburgh physics professor David Snoke has described how the use of teleological concepts and terms is widespread in contemporary systems biology, and said the scientists in that discipline therefore are actually working under a design framework rather than an evolutionary one. "Many have demanded that the intelligent design paradigm must come up with a successful, predictive, quantitative program for biology," he commented, "but it seems that such a program already exists right under our noses."[16]

In engineering and technology various parts are treated as purposive units of a purposive whole. First the purpose is defined and then the parts are organized accordingly to reach the goal. Many industries have started to look at their systems and products from the point of view of information: the medical industry is moving from a chemical-based approach towards systems biology, which means information-based operations. Polymers have long been information-poor, but now we talk about intelligent polymers. The study of biological systems and its practical branch, synthetic biology, is a study of intelligent systems. Below are some examples of language use in systems biology.

Timing and synchronization. In car manufacturing every component has to arrive at the right time and at the right place. An early step in the chain involves manufacturing special parts. A friend of mine was manufacturing catalyzers for a big European car company. He had to be alert day and night and be available in twenty minutes. The parts had to arrive at the factory when needed—no sooner, no later. The same is true in biological systems with their various timing and control systems.

Targeting. A signal must reach its target to be useful. Biological signals do not swim randomly in a cell. They have an address and material is transported to a destination, like letters sent via the post office to a particular mailbox.

Redundancy. Engineering language talks about security systems that kick in when the main system breaks. Hospitals, for instance, have reserve systems for electrical failure. The redundancy is there to make the system more robust. In biology we have redundancies built into the genetic code. We have homologous proteins. We have two kidneys, etc.—redundancy.

Adaptation. A good example is glass that gets darker with increasing light. The manufacturing of such "intelligent glass" is much more demanding than that of normal glass. But this degree of adaptability pales in comparison to what we find in living things. Biological systems are amazingly adaptable to different conditions. We can think of, for example, changes in animal behavior, physical structure, and metabolism due to adaptation to different climates, or seasonal adaptations to extremes in temperature. Another example is the ability of our body to regulate concentrations of various chemicals like salts and sugar within narrow limits.

In recent decades, researchers have sequenced the information-rich genomes of many different organisms. Each new genome sequencing produces a flood of new data. Systems biology asks, *what does it all mean? What purposes does the sequenced information serve?* All these are teleological questions. They are the kind of questions engineers ask when they are reverse-engineering something. It's design thinking.

Systems biologist Arthur Lander summarizes the field this way: "Gene regulation, intracellular signaling pathways, metabolic networks, developmental programs—the current information deluge is revealing these systems to be so complex that molecular biologists are forced to wrestle with an overtly teleological question: What purpose does all this complexity serve?"[17]

For most systems biologists the theory of evolution is of very little practical use; instead, they use design terms openly. Writing in the journal *History and Philosophy of the Life Sciences*, Pierre-Alain Braillard underscored this characteristic of the field, contrasting it with a mechanistic, non-teleological approach:

> In this paper, I examine the emerging field of systems biology and argue that some of its approaches do not fit the mechanistic framework. I present an example of what can be called design ex-

planation and show how it differs from classical mechanistic explanations. First, it is a non-causal kind of explanation that does not show how a function is produced by a mechanism but illustrates how a system's function determines its structure. Second, it points to general design principles that do not depend much on evolutionary contingency... Although some aspects of systems biology fit the mechanistic framework, explanations used by working scientists do not always correspond to the traditional definitions of mechanistic explanations provided by philosophers.... I refer to this kind of explanation as design explanation.[18]

We can close out the chapter with a few takeaways from all this:

- Information is crucial in understanding life.

- Biological information is more than its material carrier.

- The mechanisms of chemistry and evolutionary biology are insufficient to explain the information labyrinth that makes life.

- Systems biology approaches engineering science and uses the language of systems science, which is teleological.

- Explanations based on the mutation-selection mechanism of evolution are of no practical use in synthetic biology and systems biology.

- The teleological concepts used by Wilder-Smith were far ahead of their time and have become part of the normal vocabulary of systems biology.

6. Broadcaster Bias

It is summer 1994, and I am in a field of flowers on the island of *Suomenlinna* off the coast of Helsinki. I am walking toward a camera and into a stiff wind, answering questions about evolution. Later we learned that the noise of the wind spoiled the recording, so we reshot the scene the following week. It was the first job of Amigos Media, founded by four young filmmakers—Jukka Rahkonen, Harri and Ismo Paavola, and Ville Päivänsalo.

Finland's national public service broadcasting company (Yle) had promised to finance the project, introduced to Yle as *The Big Taboo of Science*. The working title was *Project Delicate*. As we would soon find out, neither title exaggerated.

Shooting continued outside Helsinki in Cultor's research center, where I served as the research director, and then near my hometown, Lahti. It was Paavola's final study project for his university in Rovaniemi, a city in Lapland near the Arctic Circle, so he had access to first-class cameras and other equipment.

One of my roles was providing the team with some international contacts, so I wrote to origin-of-life scientist Dean Kenyon at San Francisco State University to check his background information and ask his permission to use segments from an interview he had done for Access Research Network. Years before, Kenyon had written a book with Gary Steinman that argued for the unguided origin of the first life.[1] However, he changed his mind after one of his students loaned him *The Creation of Life* by A. E. Wilder-Smith.[2] The two scientists met some years later. Wilder-Smith told me that at this meeting he had asked Kenyon if his criticism of Kenyon's theory had angered Kenyon. Kenyon had replied that since the criticism was correct, why would he get angry?

After Kenyon agreed to let us use excerpts from that earlier interview, I contacted British mathematician, astronomer, and astrobiologist Chandra Wickramasinghe to check on a story I had heard about him. A colleague had told me that Wickramasinghe's life was threatened after he and Fred Hoyle took a critical view of evolution in their 1981 book, *Evolution from Space*.[3] Police took the threat so seriously that Wickramasinghe fled with his family to his native Sri Lanka. In a letter to me he wrote that the story was true. Specifically, someone had threatened to burn down Wickramasinghe's house with him in it.

Next I wrote to Siegfried Scherer, a professor of microbial ecology at the Technical University of Munich. I had gotten to know him during my stay in Zürich. He promised an interview if the producers came to Munich. So the team rented a van and drove there.

The documentary ultimately was titled *The Deep Waters of Evolution* and was ready in 1995. Amigos Ltd. also produced an English version[4] aired on Norwegian television that won an Award of Excellence at the CEVMA Film festival in London.

Yle paid the agreed sum to the producers but did not air it. The filmmakers were met with complete silence. The silence lasted almost three years before finally the documentary was shown on Good Friday of 1998. The message from Yle was clear: *This is religious propaganda*. Now, a documentary exploring what the Bible says about origins would be a fine thing, and given the Bible's massive impact on the development of Western civilization, a perfectly appropriate topic for a Yle documentary on Good Friday or any other day of the year. However, this particular documentary offered only scientific criticisms from scientists skeptical of neo-Darwinism, not arguments based on appeals to Scripture. Airing the film on Good Friday was meant, I suspect, to obscure this fact.

What's especially telling is that Yle happily airs countless nature documentaries, year round, brimming with propaganda for philosophical materialism. One finds similar fare on British and American television.

From these documentaries we are informed that nature has produced the neck of the giraffe, the farming techniques of leafcutter ants, the dance of bees, the compound eyes of flies, the sense of smell in dogs, the beauty of roses, and the big brains, upright posture, and moral and creative capacities of human beings. We are told that nature has made everything by itself, without intelligence or plan. These nature documentaries tell us this sort of thing over and over. It's an unending drumbeat. And by and large the claims are merely asserted, with at most an imaginative narrative gloss offered to sell any given evolutionary claim. This is religion, but it's the religion of evolutionary materialism, so it's OK.

In the 1980s I was listening to a seminar presentation by the research director of a Swiss pharmaceutical company at ETH. He showed the complicated chemical structure of a medicine and commented, "Every organic chemist would be shocked if they had to try to synthesize such a molecule! Luckily, Mother Nature has made it for us." Where did he have the knowledge that Mother Nature had made it? What was the mechanism she used? Of course he didn't know; he only believed it. This unconscious religiosity is all too common in the science community, and the broadcast media ensure that it's presented as scientific fact day after day.

The silver lining is that contemporary design arguments are generating so much attention and discussion in various universities around the world that Yle finally felt compelled to touch the topic. In winter 2005, they contacted me and asked me to sit for an interview. I accepted the invitation and we met in my laboratory on the university campus. I told the filmmaker for the planned news piece that he had an excellent opportunity to make a balanced and factual program digging into the arguments. Or he could choose the typical approach and make an evolution critic look like a fool. Unfortunately, he chose the latter alternative.

The program aired February 7, 2005,[5] and only one of my many comments from the interview made it into the final product. The opponents in the film had six times more screen time, and they used it to insist that evolution has been repeatedly confirmed and that the critics of evolution are anti-intellectual. Hanna Kokko, presently a professor in Zürich, informed viewers that "playing with such opinions is not intellectually honest." Another person in the film, a high school biology teacher, warned that "this thing eats away the basis of science and the basis of producing scientific knowledge." The rhetorical fallacy known as *ad hominem* attack was everywhere present, usually in close company with the fallacy of begging the question.

My loyal nemesis, Professor Leikola, also made an appearance, informing viewers that "there has always been a group of people—sometimes bigger, sometimes smaller—to whom scientific thinking is so strange that they rather believe in astrology." He neglected to mention that many astrology enthusiasts also believe in that magical thinking known as modern evolutionary theory. The adherents of these two belief systems merely overlap, of course. They are not coterminous. What evolution and astrology share in common is a penchant for stories so vague and elastic they are difficult to falsify. On the webpage of Yle the program was described as follows: "The views of many people on Darwinism and evolution seem to be on the wrong track." I totally agree – but for different reasons!

In autumn 2007 Yle showed another documentary discussing intelligent design.[6] In contrast to what was claimed on the webpage, the film was far from balanced and contained nothing new scientifically. The film described ID theory as a political movement and scared people with warnings about fundamentalism and theocracy.

The film also missed the fact that behind the birth of modern science were men like Galileo, Boyle, Faraday, Maxwell, Newton, Pascal, and Pasteur, scientists who believed in real design in nature and in a cosmic designer, a faith that inspired them to go looking for the underlying rational order of the natural world.

The documentary simply assumed evolution to be true, and it repeated the assertion *ad nauseam*—persuasion through repetitive conditioning rather than rational explanation and argument. When watching the show one got a feeling that critics of evolution were somehow dangerous. The reasons, however, remained unclear.

Public television in the United States played a similar game during this decade. Spokespersons for the lavishly promoted PBS series *Evolution* stated that "virtually all reputable scientists in the world" support neo-Darwinism, and the series producers used the claim to keep any criticism of modern evolutionary theory out of the eight-hour production.

The claim didn't go unchallenged. Discovery Institute's Center for Science and Culture (CSC) is an institutional hub for scientific criticism of modern evolutionary theory and for the development of the theory of intelligent design.[7] Partly in response to the series and this claim, Discovery Institute issued a Dissent from Darwinism list in the fall of 2001. The list (described in the previous chapter) began with one hundred names and steadily grew to many hundreds of scientists. These scientists, keep in mind, were just those Darwin skeptics who were in contact with Discovery Institute *and* willing to voice their skepticism publicly. The fact that a disproportionate number of the signers are tenured faculty members, nearing retirement, and/or emeritus faculty is what one would expect in an academic culture where voicing skepticism of Darwinian dogma can be dangerous to one's career.

The Dissent list punctured the claim that "virtually all reputable scientists in the world" support Darwinism, but the misleading claim went right on serving as cover for the PBS series' one-sided and inaccurate discussion of evolution and the Darwin/Design controversy.

Discovery Institute also published a book in 2001, *Getting the Facts Straight*, detailing the many scientific and historical inaccuracies in the series. Each of the eight episodes received its own chapter, but an executive summary encapsulated some of the key shortcomings in the series as a whole:

> We are told that "powerful evidence" for the common ancestry of all living things is the universality of the genetic code. The genetic code is the way DNA specifies the sequence of proteins in liv-

ing cells, and *Evolution* tells us that the code is the same in all living things. But the series is badly out of date. Biologists have been finding exceptions to the universality of the genetic code since 1979, and more exceptions are turning up all the time. In its eagerness to present the "underlying evidence" for Darwin's theory, *Evolution* ignores this awkward—and potentially falsifying—fact.

Evolution also claims that all animals inherited the same set of body-forming genes from their common ancestor, and that this "tiny handful of powerful genes" is now known to be the "engine of evolution." The principal evidence we are shown for this is a mutant fruit fly with legs growing out of its head. But the fly is obviously a hopeless cripple—not the forerunner of a new and better race of insects. And embryologists have known for years that the basic form of an animal's body is established before these genes do anything at all. In fact, the similarity of these genes in all types of animals is a problem for Darwinian theory: If flies and humans have the very same set of body-forming genes, why don't flies give birth to humans? The *Evolution* series doesn't breathe a word about this well-known paradox.

Most of the remaining evidence in *Evolution* shows minor changes in existing species—such as the development of antibiotic resistance in bacteria. Antibiotic resistance is indeed an important medical problem, but changes in existing species don't really help Darwin's theory. Such changes had been observed in domestic breeding for centuries before Darwin, but they had never led to new species. Darwin's theory was that the natural counterpart of this process produced not only new species, but also fundamentally new forms of organisms. *Evolution* has lots of interesting stories about scientists studying changes within existing species, but it provides no evidence that such changes lead to new species, much less to new forms of organisms. Nevertheless, it manages to give the false impression that Darwin's theory has been confirmed.[8]

The *Evolution* series subsequently was pushed in public schools, and Discovery Institute pushed back with its viewers' guide and science documentaries critical of evolutionary theory. One of these, *Unlocking the Mystery of Life*, made in partnership with a California-based film company, Illustra Media, even aired on several regional PBS stations.

Discovery and Illustra were able to take advantage of the fact that the gatekeepers at some regional PBS affiliates had a greater interest in balanced coverage of evolution than was the case at the national level. Alas, these moments of balance come all too rarely, and as in Finland, are set over against an unending stream of evolutionary propaganda.

Of course the broadcast media could treat the subject in a different way. They could ask evolution supporters tough questions about the origin of the genetic code, the origin of biological information, the origin of basic structures in the animal kingdom, the difference between microevolution and macroevolution, the pattern of abrupt appearance and stasis in the fossil record. They could ask them about evolutionary theory's failed junk DNA expectations. There are science documentaries that have done this, but rarely are any of them allowed to air widely on broadcast television.

The fate of a recent Australian-produced documentary offers a case in point. In the autumn of 2008 an Australian film team came to my office. Philosopher Tapio Puolimatka and I had promised to give an interview for their film depicting the life and Beagle voyage of Charles Darwin. Before coming to Finland the team had cruised along the South American coast and visited the Galápagos Islands. To guarantee balanced treatment of the subject, seven critics of evolution and seven evolutionists were interviewed. The resulting 2009 documentary, *Darwin: The Voyage that Shook the World*,[9] was described by a sympathetic critic at Movie Guide as "an extremely well-crafted exploration of who Charles Darwin was and what he believed," and as "one of the best-produced documentaries ever made."[10] Even pro-evolution critics who faulted the content of the film for criticizing evolutionary theory complimented its excellent production values. That is the encouraging news. The discouraging news for Finns is that although two Finnish professors were interviewed for the film, it was never aired on Finnish television.

Another example is the 2008 documentary *Expelled: No Intelligence Allowed*, starring Ben Stein, a writer, commentator, Hollywood movie actor, and speechwriter for two American presidents. Stein served as the film's onscreen narrator and interviewed both critics and well-known supporters of evolution. Several million viewers saw the film in the United States, many in theaters, and it generated considerable discussion. The pro-evolution American Association for the Advancement of Science (AAAS) called it dishonest and deceitful, designed for the insidious purpose of injecting religious ideas into public schools and science education. But the main point of the film was to tell the stories of several scientists ousted from their positions for offering scientific arguments against materialism. At least one of them didn't even offer those arguments in the workplace. He presented them in a book, written on his own time, even while maintaining an outstanding record of peer-reviewed publication in his field.

Despite the film's immense popularity, high production standards, and impressive list of people interviewed, including atheistic biologist Richard Dawkins, the film was never aired in Finland.

When watching the sort of nature programs that do regularly air in Finland, throughout Europe, and in the United States, one gets the impression that it was relatively easy for blind natural forces to make complex molecules, animals, and plants. Never mind that even the simplest living organism is far more sophisticated than anything our most advanced human engineers have managed to design. And never mind that the Darwinian explanations for these structures are habitually vague.

Man has often copied ideas from nature but never exceeded nature at its most intricate. And there is no end in sight to how much our best engineers have to learn from the biological realm; think of such things as echolocation, the orienteering skills of birds and salmon, the unbelievably complicated eyes of insects, and the ability to fly as nimbly as a dragonfly. At the time of this writing, the Denver Museum of Nature & Science had a major exhibit focused on highlighting many of them. "*Nature's Amazing Machines* uses real objects, scientific models, and fun activities to show the marvels of natural engineering," the museum website rightly gushes. Marvelous indeed!

What can we learn from these examples noted above, from the examples on display at the Denver museum, and from countless other biological engineering marvels? A whole field—biomimetics—is dedicated to seeking the answers.

Orchestrating an Animal

FOR BIOMIMETICISTS, nature's school room regularly puts on a clinic we might call "How to Build an Animal." I refer not to the building of a fundamentally new animal form, such as the first birds far in the distant past, but just to the ordinary birds-and-bees process of embryo development. We know, of course, that each animal starts as a fertilized cell that develops to an adult individual made of many cells. It's during this period of embryonic development that the basic body plan of an animal is formed. Proteins are organized into higher-level structures. Different cell types are formed. Out of them arise different tissue types. And from these arise different organ types and, finally, the fully formed animal. This complicated series of events during embryonic development cannot be explained simply by considering the origin and functioning of genes. To understand how an animal is built, we also have to understand what happens during an embryo's development.

During this phase, the right types of cells have to be made at the right times and brought to the right places. The correct amount of each is also needed. In this process some genes are expressed while others that are not needed in a certain growth phase are repressed. During the development of a human, for instance, embryonic red cells contain hemoglobin that is different from that in mature red cells. Brain cells produce enzymes involved in transmitting nerve impulses, while intestinal cells produce enzymes to degrade food in the alimentary track. These proteins function in totally different environments and have completely different tasks.

And those are just three examples. A human being has hundreds of different cell types. Each needs to be formed at the right time, in the right amount, and delivered to the right place during embryological development. If the orchestra doesn't perform exquisitely, the animal self-destructs. How could such an extraordinary orchestra, complete with its unseen conductor, have evolved one micromutation at a time? Evolutionists haven't a clue, but the dogmatic ones are certain it must have happened somehow because the alternative, intelligent design, is for them verboten.

Orphan Genes

CRUCIAL TO orchestrating embryonic growth are what are called developmental gene regulatory networks (dGRNs). An important part of these networks are proteins that bind to DNA and regulate expression of proteins and RNA. The system is unimaginably complex. Did it all evolve from a simpler common ancestor? If so, we should see evidence of that as we compare regulatory networks across different species, genera, families, and phyla. The pattern we actually find is raising eyebrows even among evolutionists. Evolutionary biologist Seirian Sumner describes the emerging problem:

> These data are telling us to put to bed the idea that all life is underlain by a common toolkit of conserved genes. Instead, we need to turn our attention to the role of genomic novelty in the evolution of phenotypic diversity and innovation.

> We can now sequence *de novo* the genomes and transcriptomes (the genes expressed at any one time/place) of any organism. We have sequence data for algae, pythons, green sea turtles, puffer fish, pied flycatchers, platypus, koala, bonobos, giant pandas, bottle-nosed dolphins, leafcutter ants, monarch butterfly, pacific oysters, leeches… the list is growing exponentially. And each new genome brings with it a suite of unique genes. Twenty percent of genes in nematodes are unique. Each lineage of ants contains about 4000 novel genes, but only 64 of these are conserved across all seven ant genomes sequenced so far.

> Many of these unique ('novel') genes are proving important in the evolution of biological innovations. Morphological differences between closely related fresh water polyps, Hydra, can be attributed to a small group of novel genes. Novel genes are emerging as im-

portant in the worker castes of bees, wasps and ants. Newt-specific genes may play a role in their amazing tissue regenerative powers.[11]

Where these novel genes come from is pure speculation. Macroevolution needs coordinated changes for new basic body plans, since only a full suite of coordinated changes could provide beneficial rather than harmful change. That's crucial, since natural selection tends to weed out harmful changes—the more harmful, the more likely they are to be quickly weeded out. And other findings have rendered that problem an even bigger headache for evolutionists.

Lethal Mutations

EVOLUTION NEEDS mutations in genes expressed in the early phases of embryonic development, since it's in the early phases that basic body plans and organs are laid down. For evolution to do anything more than tinker with existing species within narrow limits, it has to be able to mutate the fundamentals—body plans, core architecture, that sort of thing. Such early phase mutations do occur, but here's the catch: Those mutations are always harmful or lethal. The physiologist and Nobel Laureate Thomas Hunt Morgan performed systematic experiments with fruit flies (*Drosophila*) in the early twentieth century. He saw that those mutations that influence the basic structures of the animal, which occur early, are without exception harmful, leading to crippling malformations or death. (See Figure 6.1.)

FIGURE 6.1—Non-viable mutants of fruit fly *Drosophila*: From left to right: A) Mutant with twisted wings, B) Mutant with short wings that cannot fly, C) "Antennapedia" mutant with legs where antennae should be and that cannot multiply, D) "Eyeless" mutant that is blind.

Experiments carried out later with various test organisms have come to the same conclusion: Mutations that influence the basic body plan of organisms are harmful, often lethal. The geneticist John F. McDonald calls this problem "the great Darwinian paradox." He expresses the problem as follows: "Those (genetic) loci that are obviously variable within natural populations do not seem to lie at the basis of many major adaptive changes, while those loci that seemingly do constitute the foundation of many if not most major adaptive changes are not variable within natural populations."[12]

So the mutations that evolution needs in order to build new body plans do not occur, and those that do occur, evolution doesn't need.

Beyond the Gene

TODAY MORE and more biologists are convinced that there is information in the cell outside DNA. In other words, DNA does not control all cell activities but is only one of the necessary requirements for the cells, tissues, and organs to function properly. In particular, developmental biologists have observed that the formation of body plans is influenced by the form and structure of the embryonic cells, and such information is outside DNA.

We transfer to the next generation not only DNA but the whole fertilized egg cell. Scientists talk about epigenetic (outside DNA) information that the next generation can inherit.[13] Possible information carriers in the cell are, for example, properties of the cell support systems. Like amino acid- and nucleoside-based polymers, sugar-based polymers can contain complicated code systems that can be put together in many different ways. With only a few different sugars, thousands of different sequences can be built. Biologists talk about *sugar code* and compare it with DNA code.

DNA information, then, is only a portion of the total body of biological information. That is at least one reason why we can mutate DNA as much as we like and no beneficial new anatomical structure is ever formed. Research in cell biology has shown that the traditional mechanism of neo-Darwinism is incapable of producing major evolutionary changes.

So then, why not simply expand neo-Darwinism to include epigenetic mutations? Couldn't genetic plus epigenetic mutations carry us from the first single-celled organism to all the varied life we find around us on planet Earth? Stephen Meyer notes that there are at least two critical problems with that idea. For one, the structures where epigenetic information is found "are much larger than individual nucleotide bases or even stretches of DNA" so that they're "not vulnerable to alteration by many of the typical sources of mutation that affect genes such as radiation and chemical agent." Second, "to the extent that cell structures can be altered, these alterations are overwhelmingly likely to have harmful or catastrophic consequences." Meyer continues:

> Altering the cell structures in which epigenetic information inheres will likely result in embryo death or sterile offspring—for much the same reason that mutating regulatory genes or developmental gene regulatory networks also produces evolutionary dead ends. The epigenetic information provided by various cell structures is critical to body-plan development, and many aspects of embryological development depend upon the precise three-dimensional placement and location of these information-rich cell structures.[14]

So the epigenetic revolution is an exciting frontier of discovery, but it's a graveyard for any notion of gradual evolutionary change via random mutations. As for the broadcast media, don't expect them to televise the revolution anytime soon, at least not in a fashion that makes clear the acute challenge this revolution poses for evolutionary materialism.

7. The Church Evolves

In summer 2003 I was driving with my wife in southern Sweden when my mobile phone rang. The person introduced himself as Pekka Salminen, a veterinarian from the small city of Pello, near the Swedish border in Lapland. He had just read the Finnish translation of William Dembski's *Intelligent Design* and was calling to express his enthusiasm for the book. He had connections to the leadership of the Lutheran church, and he quickly went to work using the subject matter of the book to organize a seminar for the next national church conference in May 2005.

These conferences have had very controversial guests. The American Bishop John Shelby Spong drew much public discussion when he was a guest in 2003. According to Spong, Darwin's evolutionary theory has "destroyed forever the power of the traditional Christian myth."[1] Given that the church conference had invited him as the main speaker, surely they would also welcome a scholar whose arguments, if true, demonstrated that Darwin had not in fact destroyed the foundations of Christianity. Or so thought Salminen.

To Salminen's surprise the topic suggestion was met by intense opposition. The organizers had no interest in exploring intelligent design or giving a platform to any critic of Darwinism. Salminen, however, did not give up. Like the persistent widow in the parable, he wore them down and the organizers eventually accepted as a speaker intelligent design proponent and philosopher of biology Paul Nelson from the United States, with Salminen promising to pay Nelson's travel costs from his own pocket. I contacted a professor of philosophy at Helsinki University because the university's president, Ilkka Niiniluoto, had in a letter the year before told Nelson that he thought the topic of design was best suited to the philosophy department. Fine. We would try that door. Alas, there was no interest. I tried the faculty of theology and got the same negative response. Oh well, we thought. There was still the conference Nelson would be speaking at.

He arrived in Helsinki and kicked off his visit with a talk on the origin of life at a private seminar. The next day we flew to Oulu, a northern city in Finland where the church conference was being held. There we found out that Nelson's seminar was not in the main conference hall. It wasn't immediately clear where it was located. Finally we learned that it was to take place in a school building about a mile away from the actual conference location. Only then, after Nelson had flown there all the way from the United States, did they bother to inform him that he was not an official guest of the conference.

The silver lining was that a few people still managed to find their way to Nelson's talk, despite the organizers' best efforts to marginalize it. Nelson gave an excellent talk on the problem of evil in nature and in mankind. This was followed by a lively panel discussion on the question, "Can a Church That is Bound to Naturalistic Scientism Be Credible?" I was joined on the panel by microbiology professor Pentti Huovinen, physicist Markus Olin, and pastor Päivi Jussila—all theistic evolutionists.

I had earlier met Huovinen's brother, the late Lutheran bishop Eero Huovinen, when he visited the research center at Cultor where I was the director. One of my fellow scientists asked me jokingly what on earth I would discuss with the bishop. I showed the bishop the research center and told him about our sugar-related research and asked him whether he knew about my views on Darwin's theory. He said he did and that he would like to discuss them with me some time. I gave him a copy of my article "The Worldview Character of Evolution" and said that I would be happy to. The discussion, however, never took place.

Bishop Huovinen had written the text for Finland's new Evangeli-cal-Lutheran catechism, which was accepted in a 1999 church council. About creation the text says, "God is the creator of everything. With his word he has created the whole world. Science studies the riddle of the origin of the world and development of man. Faith trusts that behind all is the creative will and the love of God for his creatures."[2]

The text looks nice but in practice it surrenders the whole field of rational thinking and research to materialism. The Christian faith has with this definition been cut off from reality and moved to the realm of subjective beliefs, isolated from claims of materialistic science.

Berkeley law professor Phillip Johnson writes that the theistic evolu-tionist who takes this route gives away more than he may realize:

> He might explain his position in words like these: "Yes the diver-sity and complexity of life are the result of evolution. Yes, evolution is a blind, unsupervised and unintelligent process. Yes, we humans are the result of a purposeless and natural process that did not have us in mind. Isn't it wonderful that science (reason) has discovered all this *knowledge*? Of course none of this scientific knowledge con-tradicts my religious *belief* that God is our maker, because science is known to us by reason and religion is a matter of faith." [emphasis in original]

This separate-domains outlook is rarely stated that baldly, Johnson explains, because "clear, simple statements tend to arouse our common sense, which tells us [the theistic evolutionist] is trying to ride two horses that are going in opposite directions." Johnson adds, "It is a sophisticated mistake, and hence it has an irresistible attraction to intellectuals who are looking for a way to convince themselves that there is no need to deal with the conflict between theism and scientific naturalism."[3]

I've had dealings with other bishops in Finland. One that occurred in 1981 was equally uninspiring. The bishop in this case was Aimo T. Nikolainen and the occasion was a meeting in the Lutheran Church's training center to discuss evolution and creation. I was there with Wild-er-Smith, and I had a bad feeling right off the bat.

All of the Finns present knew English. Indeed, Lutheran bishops in Finland are, as a rule, proficient in at least Swedish and English, and the older generation is also fluent in German. And yet Nikolainen and the others there insisted on carrying out the discussion in Finnish (with me forced to interpret in both directions). This was an unusual, and an unusually rude, way to welcome a foreign guest.

Another strange feature of the meeting was that most of them did not really discuss or listen but merely read passages from their own books. Bishop Nikolainen was among those. He insisted the whole issue had long ago been settled and complained that persons like myself only disturb the excellent relations between church and university. He was convinced that opposition to evolution was based solely on misinterpretation of the Bible, with his more appropriately flexible interpretation being perfectly in line with neo-Darwinism. As best I could tell he had settled into his pro-Darwinian views with precious little understanding of the relevant science. He took the claims of the Darwinists primarily on faith.

I was not surprised, because I was not unacquainted with the type: the clergyman who will concede almost anything to the scientific materialists provided they leave some semblance of his religion intact and don't make him do any hard scientific thinking. The deal they offered, I came to realize, was much more inviting to them than what I offered. I was pressing the bishop to wrestle with the scientific evidence, which apparently was tantamount to asking him to go on an all-castor-oil diet. The more appealing offer from the evolutionists was: Let the evolutionists worry about the science; the bishop should just focus on faith and good will to man and such. Science and faith, they reassured him, are two separate, non-overlapping domains. Let the scientists do their job and they would leave the bishop to do his. Everyone would be happy.

At the conclusion of the meeting, Wilder-Smith and I left while others remained for dinner and sauna (a common feature of Finnish gatherings). Lennart Saari (see Figure 4.3, in Chapter 4) later told me that the vice-director of the training center took some of Wilder-Smith's supporters aside and urged them to back off their efforts to promote his anti-Darwinist crusade. Why? The training center director said that our eastern neighbor, the Soviet Union, did not view our activities favorably. That much was surely true, since Wilder-Smith's books were considered anti-communist and were regularly smuggled into the Soviet Union through Finland.

Christian literature and books critical of Darwinism were both forbidden in the Soviet bloc countries. But Wilder-Smith was eager to reach people trapped inside the Soviet propaganda bubble, so he travelled behind the Iron Curtain to East Germany and Poland, despite being warned by the Swiss police that it was dangerous. The meetings set up during his visit could not be advertised, but the news spread by word of mouth and the meeting rooms were packed. Wilder-Smith later told me that he wondered why so many of the audience members at these meetings carried family albums with them—until he realized that the albums were filled with photocopies of his books.

I greatly admired Wilder-Smith's courage in traveling behind the Iron Curtain, much preferring it to the go-along-to-get-along attitude taken by the Lutheran training center director eager not to ruffle any Soviet feathers. My own view of Christian faith is that one of its main purposes is to disturb established institutions with sharp questions and function as their conscience. As a young Christian who criticized naturalistic Darwinism, I was surprised that the representatives of the Lutheran Church, instead of supporting and protecting such efforts, often opposed them.

With Friends Like These...

As I got to know more and more Darwin skeptics from Europe and the United States, I learned that they too had a clergy problem—not with all pastors and theologians, of course, but with a significant and sometimes vocal minority. The pro-evolution lobbying group called the National Center for Science Education (NCSE) works tirelessly to persuade religious Americans, Christians especially, that Darwinism is good medicine, and they get plenty of help from certain religious figures. John West, associate director of the Discovery Institute's Center for Science and Culture, commented on their work in a 2009 article:

> On a taxpayer-funded website that the NCSE helped design, teachers and students are directed to a list of statements by religious groups endorsing evolution, and Eugenie Scott, the group's executive director, encourages biology teachers to spend class time having students read statements by religious leaders supporting evolution. Scott even suggests that students be assigned to interview local ministers about their views on evolution—but not if the community is "conservative Christian," because then the lesson that "Evolution is OK!" may not come through….
>
> The NCSE also encourages inviting ministers to testify before school boards in favor of evolution, and it has created a curriculum to promote evolution in the churches. The NCSE even has a "Faith Network Director" who claims that "Darwin's theory of evolution… has, for those open to the possibilities, expanded our notions of God." Other evolutionists have collected signatures from liberal clergy in support of evolution as part of "The Clergy Letter Project" and have urged churches to celebrate "Evolution Sunday" on the Sunday closest to Darwin's birthday.[4]

Why the lavish campaign? "This attempt to put a religious face on modern evolutionary theory is an effort to deal with what might be called Darwinism's 'Dawkins' problem," West goes on to explain. "Oxford biologist Richard Dawkins is one of the world's foremost boosters of Darwinian evolution. Unfortunately for evolutionists, Dawkins zealously expounds the anti-religious implications of the theory, and regularly denounces religion…. By highlighting the religious defenders of evolution, the NCSE undoubtedly hopes to depict Dawkins as a fringe figure whose views are not representative of Darwinists as a whole."

But the numbers tell a different story. "According to a poll of scientists listed in *American Men and Women of Science*, 57.5% of the biologists who responded were atheists or agnostics and 59.4% disbelieved or were agnostic about personal immortality," West writes. "The nation's most elite biologists are even more atheistic. According to a 1998 survey of members of the National Academy of Sciences (NAS), 94.4% of the NAS biologists are atheists or agnostics. A similar percentage rejects life after death."[5]

Another, more recent survey polled Americans generally and found that "43% of Americans now agree that 'evolution shows that no living thing is more important than any other,' and 45% of Americans believe that 'evolution shows that human beings are not fundamentally different from other animals.'" This, of course, clashes with the Judeo-Christian belief that humans are made in the image of God and thereby possess inherent dignity and rights.

The survey also found evidence that the theory of evolution is reshaping people's understanding of morality, with 55% of Americans now contending that "evolution shows that moral beliefs evolve over time based on their survival value in various times and places."[6]

Sadly, the religious leaders who have rushed to make their peace with Darwinism are either blissfully unaware of these trends, willfully heedless, or actively complicit.

Even some secular thinkers find themselves bemused by such behavior. In 1982 Wilder-Smith and I were at the Federal Institute of Technology in Zürich (ETH) for a seminar Paul Feyerabend had helped organize. Feyerabend was a prominent philosopher of science and a professor at the University of California-Berkeley who described himself as a nihilist. Later, in his book *Farewell to Reason*, he commented on his experience at the seminar. "In 1982 Christian Thomas and I organized a seminar at the Federal Institute of Technology in Zürich with the purpose of discussing how the rise of the sciences had influenced the major religions and other traditional forms of thought," he wrote. "What surprised us was the fearful restraint [with?] which Catholic and Protestant theologians treated the matter—there was no criticism either of particular scientific achievements or of the scientific ideology as a whole." That was in a footnote. His comments in the body of the text were even more pointed:

It is a pity that the Church of today, frightened by the universal noise made by scientific wolves, prefers to howl with them instead of trying to teach them some manners... When I was a student I revered the sciences and mocked religion... I am surprised to find how many dignitaries of the Church take seriously the superficial arguments I and my friends once used, and how ready they are to reduce their faith accordingly. In this they treat the sciences as if they, too, formed a Church.[7]

His description of the "Church of today" well applies to a late archbishop of the Finnish Lutheran church, Mikko Juva. After Wilder-Smith's visit, the archbishop wrote in Finland's largest newspaper, *Helsingin Sanomat*, "There is no real contradiction between church theology and academic biology, hardly even a problem. Without doubt the living nature has gradually developed from primitive life forms. It seems very probable, too, that evolution occurred mainly the way Darwin pictured it..."[8]

It's all the more disappointing that Juva took this accommodationist view, given that he had previously studied how philosophical naturalism came to dominate Finland. It "occurred actually in a few years during 1883–1885, a period brimming with spirited activity and battle," he wrote in his published work on the subject. "At first it seems strange that, especially in our remote country, affected relatively late by European cultural influences, such a break was expressed so sharply and in such a short period."[9] Naturalism's takeover, he went on to explain, began first in the southwest corner of Finland in Turku and spread from there to the rest of the country, to universities and finally to theological teaching. Surely he had to see that it swept through on the back of Darwin's theory of evolution, which was just then taking Europe by storm.

The previous bishop of my own city of Espoo, Mikko Heikka, is another in this mold. He sees no contradiction between the Christian worldview and neo-Darwinism, noting that contemporary evolutionary biology makes room for the selfish and social/ethical aspects of human behavior, much as Christianity does, and thus, "The evolutionary and

FIGURE 7.1—Bishop Mikko Heikka.

Christian view of man are not far from each other…. Both Huxley and Shaftesbury are within the Christian tradition."[10] And a year before he made that statement, he wrote in a Finnish magazine that "church people are rather happy than sad over Charles Darwin's discovery."[11]

By "church people" Heikka apparently means those who have fully swallowed naturalism, since he can't be referring to the many Christians who are critical of Darwinism and understand its devastating influences. Using Heikka's logic one could as well say that church people are happy about the ideas of Karl Marx because they explain both compassion for the worker and the conflict we find between rich and poor, much as Christianity offers explanations for both—never mind that Communism denies some of the core teachings of Christianity. Soviet leader Josef Stalin, incidentally, attended ecclesiastical school as a boy but gave up his faith after reading Darwin. He then used Darwin to draw others into atheism. In his biography of Stalin, E. Yaroslavsky relates an incident shared by G. Glurdjidze, one of Stalin's boyhood friends:

"I began to speak of God. Joseph heard me out, and after a moment's silence said:

"'You know, they are fooling us, there is no God….'

"I was astonished at these words. I had never heard anything like it before.

"'How can you say such things, Soso?' I exclaimed.

"'I'll lend you a book to read; it will show you that the world and all living things are quite different from what you imagine, and all this talk about God is sheer nonsense,' Joseph said.

"'What book is that?' I enquired.

"'Darwin. You must read it,' Joseph impressed on me."[12]

Many others have lost their faith after learning about evolution. The late professor William Provine is an example. This is how he described the process:

I was a Christian but I never heard anything about evolution because it was illegal to teach it in Tennessee.... [Provine's college professor] started talking about evolution as if it had no design in it whatsoever. And I came up to him, and I said, "You left out the most important part." And he said, if you feel the same way at the end of one quarter, I want you to stand up in front of the students in this class and tell them this deep lack in evolution. And I read that book so carefully; I could find no sign of there being any design whatsoever in evolution. And I immediately began to doubt the existence of a deity. But it starts by giving up an active deity, then it gives up the hope that there's any life after death. When you give those two up the rest of it follows fairly easily. You give up the hope that there's an immanent morality. And finally, there's no human free will. If you believe in evolution, you can't hope for there being any free will. There's no hope whatsoever of there being any deep meaning in human life. We live, we die, and we're gone. We're absolutely gone when we die.[13]

Provine understood perfectly well that modern evolutionary theory is inimical to orthodox Christianity, and he is not alone in this. In 1998 a professor friend asked me to talk at a natural philosophy society meeting. My presentation was titled "A Skeptic Evaluates the Theory of Evolution." In the audience was a well-known Marxist atheist. After I mentioned that I had been prevented from speaking in some churches thanks to my skeptical view of evolutionary theory, the Marxist commented that "the church is crazier than I thought." He, like Provine, understood perfectly well that mainstream evolutionary theory is no fellow-traveler of Christianity, and instead is squarely at odds with Christianity's vision of reality as richly teleological.

Trust and Obey... Darwin?

The Finnish Bible Institute has long been a fortress of biblical Christianity, so when I spoke there in the mid 1990s about the evidence for design, I was surprised when the rector of the institute said that in these matters he rather listens to "scientists." With that statement he placed me outside science in spite of my extensive track record as a scientist. Later, when a Christian magazine reported about the event, my name was not even mentioned; only the other speaker, a theistic evolutionist, was named.

The rector was no outlier. Eero Junkkaala is a theologian and archeologist long affiliated with the Bible Institute. His book, *In the Beginning God Created: Faith in Creation and the Scientific Worldview,*[14] gives the whole playground of reason to the materialists. He does not seem to understand the influence of paradigms on human thinking.

FIGURE 7.2—I am speaking alongside Eero Junkkaala at a Christian youth conference in August 1998.

Unlike the founders of modern science—such as Francis Bacon, Robert Boyle, and Isaac Newton—Junkkaala has an overly optimistic and non-critical view of human reason in the interpretation of facts. This naiveté, combined with his not being a scientist, leads him to simply trust the scientific "authorities." Among them are atheist biologist Jerry Coyne and theistic evolutionist Denis Alexander. Both agree with Dawkins that living things look like they were designed but that design does not belong in the explanatory toolkit of origins science. Junkkaala accepts the opinions of these authors without criticism, but accuses evolution's critics of lying and closing their eyes to the evidence.

His bias also leads him to recycle claims easily exposed as bogus with five minutes of research on the internet. According to Junkkaala, proponents of intelligent design have published practically no peer-reviewed articles in support of the design position, but in fact, design theorists have published scores of such papers.[15] If his book is any indication, he has not carefully read a single one of these.

Indeed, his reference list contains practically no references to the writings of leading proponents of intelligent design—a group that includes but is not limited to Douglas Axe, Stephen Meyer, Michael Behe, Robert Marks, William Dembski, Wolf-Ekkehard Lönnig, John Sanford, Michael Denton, Jonathan Wells, Scott Minnich, Branko Kozulic, David Snoke, Jed Macosko, Russell W. Carlson, Paul Chien, Colin Reeves, David Able, and Richard von Sternberg. Their work has been recognized even by many evolutionists.

Peer Review or Peer Pressure?

MORE FUNDAMENTALLY, Junkkaala's faith in the peer-review process itself is questionable. The noted cosmologist and mathematical physicist Frank Tipler has called the system into question,[16] and he has plenty of company. As one study of peer review concluded, "In its current form, peer review offers few incentives for impartial reviewing efforts."[17]

Note that I have little reason personally to fault the system of peer review. I've thrived under it. Scientists are typically rated based on their number of peer-reviewed papers and by how many times other scientists have cited their work. By this yardstick I have done reasonably well, with about 140 reviewed papers and over 5,000 citations, and this despite spending nine years in private industry. It's a fine track record of peer-reviewed publication; however, that does not make me a fine scientist. Often the review was quite superficial, and those times when I tried to publish unconventional results, the reviewers turned the paper down. In hindsight, those unconventional results involved some of the most interesting science I did! In other words, the more groundbreaking the science I did, the harder it became to get the work accepted.

And when we did manage to get such work circulated in the research community, it was frequently greeted with hostility and a knee-jerk assumption that our experiment must have been conducted ineptly. For instance, in 1985 my team discovered that a peroxidase enzyme catalyzes the opening of an aromatic ring. If that sounds like scientific gobbledy-gook to you, all you need to grasp here is that the results were surprising and outside the stream of conventional wisdom on this particular matter. But we had been very careful in how we ran the experiment, so we went ahead and shared the results in a science meeting in Vancouver. The reactions were swift. "Peroxidases do not open aromatic rings." And "Your enzyme is not pure." And "Your analysis is wrong." But as it so happened, our results were correct and we later published them in a prestigious science journal.[18] The results were later replicated many times.[19]

In 1986 one of my doctoral students showed that lignin peroxidase mainly polymerizes lignin.[20] A competing group had just published the opposite result. I got a letter from a member of that group saying he admired our work, since we published what we saw while they had published what they hoped to see.

In 1987 I told a British professor of biochemistry that we were trying to crystallize a lignin peroxidase. "It is a glycoprotein and glycoproteins do not crystallize easily; no use trying," was the feedback. But it crystallized anyway.[21] In 1999 my team showed that a well-known commercial enzyme has a number of side activities. We submitted a paper reporting on them to a biochemical journal, but it was turned down. The result was probably wrong, we were told. "Have you done the proper blanks?" Failing to do "the proper blanks"—that is, failing to check that the result was not an artifact of something other than enzyme activity—would have been an elementary mistake, and we were an obviously experienced research team. Our results, again, turned out to be correct.

In 2005, Mary Schweitzer published her now-famous results of soft tissues in *T. rex* bones.[22] *Discover* magazine did a retrospective, and the article's subtitle nicely captured the reaction to her groundbreaking discovery: "When this shy paleontologist found soft, fresh-looking tissue inside a *T. rex* femur, she erased a line between past and present. Then all hell broke loose."[23] Conventional wisdom said there was no way dinosaur fossils could contain soft dinosaur tissue. In the same article, Schweitzer described her experience: "I had one reviewer tell me that he didn't care what the data said, he knew that what I was finding wasn't possible. I wrote back and said, 'Well, what data would convince you?' And he said, 'None.'"

I have spoken with many other scientists with similar tales to tell. Given all this, it's no wonder that geologist Warren Hamilton takes such a cynical line on peer review. "Then as now, peer review can represent the tyranny of the majority," he writes. "I have run the peer-review gauntlet perhaps a hundred times. My papers describing and interpreting geology in more or less conventional terms have progressed smoothly, whereas publication of my manuscripts challenging accepted concepts has often been impeded, and occasionally blocked."[24]

R. L. Armstrong agrees. "In science this is an old story," he writes, "likely to be repeated again, as the defenders of common wisdom are seldom treated with the same skepticism as the challengers of the status quo... In science, conventional wisdom is difficult to overturn."[25]

Günther Blobel, winner of the Nobel Prize in physiology and medicine, put it bluntly: "Your grants and papers are rejected because some stupid reviewer rejected them for dogmatic adherence to old ideas."[26]

If peer review exerts this much pressure to conform to conventional wisdom in the sort of cases described above, how much more can we expect reviewers' "dogmatic adherence to old ideas" to block papers that explicitly undermine Darwinism and, in some cases, explicitly support intelligent design? Again, it's a wonder any such papers have made it through the peer review process.

Earlier we saw how the results of the ENCODE project showed that most DNA is functional, and how the results were attacked and dismissed for undermining the conventional, neo-Darwinian view on this point. In January 2016 another interesting episode occurred when a paper published in *PLOS ONE*, exploring the exquisite architecture of the human hand, was withdrawn after complaints that the paper, in essence, broke faith with methodological materialism. Later investigation suggested that the Chinese authors of the article used the term "creator" only to refer to the creative powers of nature, not to God. But no matter: the paper was deemed radioactive. In a retraction notice on their website, the journal editors offered this explanation for the about-face:

> Following publication, readers raised concerns about language in the article that makes references to a 'Creator,' and about the overall rationale and findings of the study.
>
> Upon receiving these concerns, the *PLOS ONE* editors have carried out an evaluation of the manuscript and the pre-publication process, and they sought further advice on the work from experts in the editorial board. This evaluation confirmed concerns with the scientific rationale, presentation and language, which were not adequately addressed during peer review.
>
> Consequently, the *PLOS ONE* editors consider that the work cannot be relied upon and retract this publication.
>
> The editors apologize to readers for the inappropriate language in the article and the errors during the evaluation process.[27]

The notice carried the word "Retraction" in bright red at the top alongside an exclamation point inside a red triangle punctuating it, as if to communicate with all due urgency: *Danger: Heterodox Ideas Ahead!*

Peer review, understand, isn't an utter waste. It functions reasonably well in correcting clear mistakes. But referees can be motivated by ideological concerns and personal interests. Peer review is especially prone to slow the progress of science when experimental results and insights conflict with the reigning scientific paradigm in a particular field.

This chapter, notice, started by considering one kind of priesthood, roughly speaking—the theologians and clergy who have made it their mission to help enforce Darwinian orthodoxy. Then the chapter moved on to talking about another kind of priesthood—those in the scientific community who use peer review to guard the current scientific orthodoxies. Both priesthoods are bad for scientific progress. Science doesn't progress by simply trusting the "authorities." It doesn't progress by using peer review to enforce orthodoxy. It progresses by following the evidence wherever it leads, no holds barred.

8. "Rationalists" Behaving Irrationally

I was in Takamatsu, Japan, tossing around in bed thanks to a seven-hour time difference. Giving up, I opened the e-mail on my notebook computer and found a message from pastor Sammeli Juntunen, from a town in eastern Finland called Savonlinna, famous for its annual opera festival. He had just finished reading a book by a Finnish philosopher, Professor Tapio Puolimatka, dealing with evolution. Puolimatka has two doctoral degrees, one in practical philosophy and the other in educational science, and he did postdoctoral work alongside the famous Christian philosopher Alvin Plantinga at Notre Dame University. Pastor Juntunen explained to me that he considered Puolimatka's book so important that he wanted to arrange a debate around it.

I promised to be available if the debate came together. Jussi K. Niemelä, the chairman of the Finnish Association of Skeptics, mathematician Virpi Kauko, vice-chairman of Finland's Darwin Society, and Professor Puolimatka agreed to participate, and the debate was set for March 13, 2009. Juntunen reserved the big Savonlinna hall for the discussion, and the local newspaper promised to finance the meeting. For the day following the debate, a lecture series was agreed upon where each panelist was to speak on a different topic:

+ Tapio Puolimatka: "Has Science Abolished God?"

+ Matti Leisola: "How Does Darwinism Function on a Molecular Level?"

- Jussi K. Niemelä: "Why Did Puolimatka Get the Pseudoscience Award?"
- Virpi Kauko: "Are Links Missing?"

Everything was set, but the organizers were in for a shock. About ten days before the agreed date, both the skeptics group and the Darwin Society representatives backed out. Puolimatka's book, *Faith, Science and Evolution*, had been published in autumn 2008, and one of Finland's internationally best known and most cited scientists had endorsed the book and recommended it to those interested in science. Despite this, the Association of Skeptics quickly slapped the publisher with a 2008 "pseudoscience award,"[1] and now they couldn't even be bothered to keep their commitment to show up and debate the arguments of the book.

On the wall of my office used to hang a text that told about the pseudoscience award they gave my Bioprocess Engineering lab in 2004. It read:

> The reason for giving the prize is the seminar organized on Friday 22 October, 2004 in the lecture Hall of Helsinki University of Technology. The seminar was organized by the professor of bioprocess engineering, Matti Leisola, and it was advertised on the webpage of the laboratory.
>
> The theory of "intelligent design" is a doctrine that criticizes the theory of evolution and the sciences that support it. To replace these, ID offers a concept of supernatural design claiming that it is an in-

FIGURE 8.1—The discussion ended up going forward without representatives from the skeptics group or the Darwin Society. Pictured, left to right: professor and philosopher Tapio Puolimatka, pastor Sammeli Juntunen, professor Matti Leisola, and as chairman, Teuvo Riikonen, rector of Savonlinna Christian Institute.

disputable consequence of scientific observations. Unlike in sciences ID uses as arguments things that are not known and when necessary forgets well known facts. The world's leading science organization, the American Association for the Advancement of Science, has in one of its board's official announcements said that there is no scientific support for the claims of ID-creationism and it should not be linked as part of science education and teaching.

With this year's pseudoscience award, the Skeptics Society wants to draw attention to the difference between science and pseudoscience, and reminds that the latter has no place in university teaching. ID-creationism has in many countries tried systematically to get a foothold in the academic world and even bring in "intelligent design" as part of biology teaching. The Association of Skeptics does not of course want to limit scientific discussion or the criticism that is part of it. To present ID-doctrine as a serious theory in a scientific or technological framework can be, however, compared to astrology as part of academic astronomy education or alchemy as part of chemistry education.

The announcement is a potpourri of misinformation, from the subtly slanted to the overtly mistaken. ID theory does not criticize the natural sciences. It uses evidences and methods from the natural sciences to critique modern evolutionary theory and scientific materialism, and to argue that intelligent design is the best explanation for certain patterns in nature. It does so based not on what is unknown but on our uniform experience in the present of what does and doesn't cause things like information and irreducibly complex machines. Thus it is based on what we know about the cause-and-effect structure of the world.

Also, in making a design inference in biology, design theory doesn't infer whether the design is natural or supernatural. There are modes of reasoning and evidence that could bear on that question, but that isn't the purview of ID.

As for what ID is most similar to, it's a historical science of design detection, as are SETI, archaeology, and cryptography.

The Finnish Association of Skeptics wasn't defending science. It was defending dogma, the dogma of evolutionary materialism. Its members tend to (a) fixate on religious topics and avoid scientific arguments whenever evolution is being questioned, (b) erect a strawman of intelligent design, and (c) invoke authority—in the above case, a statement by the AAAS. But any good scientist knows that scientific controversies can only be satisfactorily settled by evidence and careful reasoning about the evidence, not by changing the subject, invoking authority, and mischaracterizing an opponent's argument. Science progresses by confirming or overturning the accepted view on a matter through evidence and careful reasoning, not by bowing to authority and caricaturing opponents.

By ridiculing me, Finland's skeptics had placed me in good company. In a separate press release in 1989,[2] they singled out a book by A. E. Wilder-Smith in order to label it pseudoscience. I've touched on Wilder-Smith's impressive scientific credentials already, but I only scratched the surface. During my career as a scientist I have met Nobel laureates and hundreds of scientists from various fields, but none has had such a deep influence on me with his scientific versatility and personality as Wilder-Smith. He worked as a professor of pharmacology in the universities of Bergen, Chicago, and Ankara. He wrote three doctoral theses—one at Reading University about optically active ketones; one at the University of Geneva about chemotherapy for tuberculosis and leprosy; and one at ETH-Zürich on chemotherapy for mycobacterial diseases. He was a research director and a consultant of a Swiss pharmaceutical company from 1950 to 1960, wrote over fifty scientific papers, was involved in about 300 patents, and for many years served as a consultant on drug problems for NATO forces in Europe. He also was chosen in four consecutive years as the teacher of the year and obtained three times the Golden Apple teacher's award from the medical faculty at the University of Illinois Medical Center, College of Pharmacy. To merely dismiss his work as pseudoscience is arrogant and defamatory. The appropriate response to a scientist of his stature would have been to engage the anti-evolution arguments in his book thoughtfully and honestly. Only if Darwinism has calcified into an anti-rational dogma does reflexive name-calling and dismissal make a perverse sense.

I wrote in 2005, in Finland's chemistry journal, "I have challenged the skeptics to a public debate about the origin of life. I am also gladly involved in organizing a symposium to discuss the scientific credibility of Darwinism and the philosophical nature of modern science."[3] In the years since then, no one has taken me up on the offer.

Professor Valtaoja offered the following explanation: "We scoundrels avoid open discussion because we are not interested to argue about whether the earth is flat simply because the Bible says so. This is what creationism is all about."[4] Actually, neither the Bible nor creationism teaches that the earth is flat. The theory of intelligent design is even a further remove from Valtaoja's caricature, since it focuses strictly on the scientific evidence and the question of design in nature, not on interpreting the Bible or bringing biblical evidence to bear on the question of origins. This is why even a non-religious figure such as the world-famous philosopher Antony Flew embraced the design argument for the origin of DNA even while remaining skeptical of Christianity. I had hoped that his recognizing the evidence for design in biology would open the door for him to study the evidence for Christianity and the Bible, and eventually embrace these as well, but that would have involved him grappling with, and accepting, additional evidence and arguments outside the scope of intelligent design, and, alas, he died not long after embracing the design argument in biology. What this highly respected thinker wasn't doing when he studied and eventually embraced intelligent design was swallowing easily refuted notions of the flat-earth variety.

Medieval scholars, by the way, didn't even believe in a flat earth. Their cosmology held to a round earth. This is why Dante's famous medieval Italian work, *The Inferno*, has the protagonist going on an imaginary journey to the center of the earth and then continuing straight through to the other side of the planet, where he comes out onto the earth's surface. Dante wasn't offering a breakthrough idea here. He was simply employing the conventional view of a round earth. The notion that medieval thinkers believed in a flat earth is an invention of secular enlightenment thinkers—a historical myth about the middle ages that self-styled "skeptics" cling to with a childlike faith despite the abundant historical evidence to the contrary.

It puts me in mind of a comic strip from Hagar the Horrible that I often use to illustrate the power of wishful thinking. Hagar and his wife are at home arguing about the soup she just made for him. Hagar says there's a fly in the soup. She insists that, no, it's a raisin. After a fierce argument, the tiny object in question flies away. "Ah-ha!" Hagar shouts triumphantly. Hagar's wife shrugs. "I'll be darned," she says. "A flying raisin." Hagar's wife didn't want to believe she had served soup with a fly in it. Scientific materialists don't want to acknowledge any fact that undermines their materialism.

Bully for Darwin (Actually, Several Bullies for Darwin)

In 2001 I received a letter from a group that had learned that I was skeptical of modern evolutionary theory. They wanted to know if I would be willing to add my name to the statement mentioned previously in these pages: "We are skeptical of claims for the ability of random mutations and natural selection to account for the complexity of life. Careful examination of the evidence for Darwinian theory should be encouraged." I was happy to respond in the affirmative, and a list with a hundred names was soon published. Most of the signers had doctoral degrees in science, with a handful having Ph.Ds. in fields that, while not part of the natural or life sciences, gave them a valuable and relevant perspective on the evolution question—e.g., engineering and mathematics. Today close to a thousand have signed this Dissent from Darwinism statement.[5] Its purpose is to show that there are serious scientists who question Darwin's theory. I am confident, incidentally, that the number of names on the list only scratches the surface, since I know scientists who, while skeptical of modern Darwinism, have not signed the document because they are afraid of the consequences.

The danger is far from imaginary. Immediately after the list became public, I received an email from the late Skip Evans of a pro-evolution lobbying group in the United States, the National Center for Science Education (NCSE). Evans was a militant atheist and defender of evolution. He wanted to clarify my motives for signing and inquired if I understood the kind of dangerous people with whom I was in contact. I knew too many good things about the scientists and scholars who had started the dissent list, and too much off-putting stuff about the pro-evolution NCSE, to be impressed by his warnings. However, I could imagine a scientist with little knowledge of either group being taken in by the NCSE's well-poisoning campaign.

Since 2010 I have been on the advisory board of the German scholarly association *Studiengemeinschaft Wort und Wissen*, and I have spoken twice (in 2009 and 2014) at their annual main conference. In the invitation letter to join their advisory board, they were considerate enough to warn me of a potential negative consequence. "In this connection we want to inform you that there is in Germany a very active group of confessing atheistic evolutionary biologists who carefully follow every move of our organization and who do not avoid ad hominem attacks," the letter explained. "It is possible that as a member of the scientific advisory board you personally can become a target. You should take this into consideration when you think about your participation." The group had no motivation to exaggerate about this. They were, after all, hoping I would join their advisory board. It was purely out of a sense of fair play that they warned me at all.

Their warning, of course, did not shock me, since I had long been a target of the evolutionary materialists. I have already mentioned several such examples in these pages. Here's another: I applied for an assistant professorship in biochemistry in 1984 at Helsinki University of Technology (TKK). I later learned from the biochemistry professor who had recommended me for the position that in the meeting of the professors' council where the hiring decision was made, one professor stood up and strongly opposed my nomination. A person so badly mistaken about biological origins cannot be a teacher in this university, he insisted. My former professor told me that he had to defend my application, telling the others in the meeting, "We are not here to discuss Leisola's worldview but his competence in biochemistry." Most of the other professors, he said, clearly felt uneasy in the situation and were looking at the walls.

In 1987 I was a consultant for the Finnish Sugar company. (In 1989 the name was changed to Cultor.) I worked at the company first as a senior scientist, then as a department manager, and beginning in 1991, as a research director. Later I heard that another consultant of the company had advised the executive director not to hire me due to my questionable views on origins and my involvement in Christian student ministry.[6]

Wilder-Smith had similar stories to relate. He told me how a Professor Hoimar von Ditfurth tried to intimidate him by contacting universities where he had obtained his degrees to expose Wilder-Smith's supposed deception. He was convinced that no one could get three doctoral degrees in such a short time and at the same time become a Fellow of the Royal Society of Chemistry (FRSC). Reading and Geneva confirmed the degrees, but the Swiss Federal Institute of Technology (ETH) did not find any information in their files of such a person. Ditfurth wrote to Wilder-Smith that he had exposed his deceit. "You have not obtained a doctoral degree at ETH!" he wrote.

Wilder-Smith called ETH and they immediately found him in their records with his correct name. Ditfurth had misspelled Wilder-Smith's name when he made the inquiry. ETH wrote a letter to Ditfurth and explained the situation. Next Ditfurth contacted the Royal Society of Chemistry inquiring into the authenticity of Wilder-Smith's title of FRSC. The society was not impressed with the inquiry and did not even care to answer and passed the letter on to Wilder-Smith.

Ditfurth wasn't the only evolutionist who tried to play this game. After an Oxford Union debate, Richard Dawkins let it be known that no one by the name of Wilder-Smith had studied at Oxford University and graduated from there.[7] In reality, Wilder-Smith had studied at Oxford from 1933 to 1935 and finalized his doctoral degree at Reading. If

Dawkins had taken only a little care in his investigation of the matter, he might have discovered this.

These incidents suggest how headlong some people can be in their efforts to discredit scientists skeptical of Darwinism.

And by the way, Dawkins wasn't finished sliming Wilder-Smith. All the information concerning the debate between Dawkins and Wilder-Smith has been lost from Oxford Union's files. When Dawkins was asked about the debate in May 2003, he admitted that the debate had taken place but then added, "Wilder-Smith I remember as a genial old buffoon... I am not interested in following up Wilder-Smith's history. The man is too unimportant to waste time over... Wilder-Smith's account lies somewhere between fantasy, lies, and paranoid delusion."[8]

Venomous much? The tirade is all the more shameful in light of this: Although Dawkins is a gifted writer and popularizer, Wilder-Smith's lasting contributions in the field of experimental biology (see above) dwarf those of Richard Dawkins. (A review in the journal *Nature* of Dawkins's career autobiography describes the man as a gifted "lyricist" but adds that "a curious stasis underlies his thought, with his view of the genome "grounded in 1970s assumptions.""[9]) Dawkins' heavy-handed dismissal fits very well his style. Dawkins is, after all, the one who used these labels for evolution skeptics, including undergraduates: "little fool," "pathetic little idiot,"[10] "ignorant," "stupid," "insane," and "wicked."[11]

Sometimes the attacks go well beyond words. I am just now reading an official report[12] by a U.S. congressional subcommittee that investigated the treatment of evolutionary biologist Richard von Sternberg. (See Figure 8.2.) Sternberg has two doctoral degrees—one in evolutionary biology and another in theoretical biology. He worked in the National Center for Biotechnology Information of the National Institutes of Health and in the Smithsonian Institution's National Museum of Natural History. He also was the editor-in-chief of a science journal published by the Smithsonian, the *Proceedings of the Biological Society of Washington*. One of his responsibilities was to sort submitted papers and send them to two or three experts for peer review.

FIGURE 8.2—Evolutionary biologist Richard von Sternberg.

As is normally the case with peer-reviewed science journals, these reviews happen anonymously. The author does not know who the reviewers are and, based on their report, the editor makes a decision on publication. The decision may be acceptance, a call for minor or major corrections, or rejection.

Stephen Meyer submitted a paper, "The Origin of Biological Information and the Higher Taxonomic Categories," which considered intelligent design as a possible explanation for the Cambrian explosion, a geological period in which a large number of basic animal body types (not just new species but whole new phyla) appeared abruptly. The three peer reviewers read the paper and unanimously favored its publication. Sternberg accepted the manuscript and it was published.

Then the persecution started.

The hue and cry went up: Science and religion had been mixed! If the paper wasn't immediately retracted, the reputation of the world-renowned Smithsonian Institution would be forever tarnished! Together with the pro-evolution NCSE, the Smithsonian hatched a plan to destroy Sternberg's career. As the congressional report details, in an early stage of this campaign, Sternberg's friends were questioned and false rumors were spread both inside and outside the Smithsonian. The rumors grew so wild that finally a colleague of Sternberg sent Sternberg's *curriculum vitae* to members of the Smithsonian as a proof of his impressive record of scientific accomplishment. Meanwhile, those gunning for Sternberg insisted that the reviewers must have been know-nothing supporters of intelligent design. Sternberg's religious motives were also questioned and his privileges narrowed. His keys were taken away, he was transferred to a much inferior office space, and he was denied access to scientific samples. The atmosphere became so hostile that Sternberg eventually decided to leave the Smithsonian.

At that point Sternberg's professional career seemed all but ruined. Who would hire such a suspicious person? Two official investigations were made and all accusations were shown to be groundless and the rumors unfounded, but no one involved at the Smithsonian corrected the rumors or apologized. (Sternberg describes on his webpage the drama of those days.[13]) In the midst of these events, Sternberg came to Finland and faced similar turmoil. The email discussion I mentioned in Chapter 3, the one that unfolded on the university's professor list, referred to the incident at the Smithsonian and it was one reason given for canceling the ID seminar.

Later I got a telephone call from one of Sternberg's friends, who asked if I could offer Sternberg a job in my laboratory till the situation calmed down. I promised to take him onto my team, but he found another job in the United States.

This story places in a whole new light the charge that intelligent design researchers are not legitimate because they do not publish their work in peer-reviewed science journals. In fact, they have had several such articles published in peer-reviewed journals. But is it any wonder it doesn't happen more often, given what happened to Richard Sternberg?[14]

Many science journals will under no condition publish a paper that explicitly makes a case for intelligent design. And many journal editors who might have considered doing so will think twice after seeing what

Sternberg went through. This was undoubtedly a key reason why so much energy was poured into harassing Sternberg. Darwinists didn't just want to punish him for heterodoxy. They wanted to make an example of him.

Junk DNA as Junk Science

THE DARWINISTS were wrong to harass Sternberg, but they weren't wrong to consider him a threat to Darwinism. Indeed, his decision to publish an ID-friendly paper isn't the half of it. Sternberg's concept of the genome as a complex information system represents an existential threat to modern evolutionary theory in general, and to a particular prediction of Darwinian thinking involving so-called junk DNA.

We have touched on the idea of junk DNA in earlier chapters, but only briefly. Let's unpack it a bit here to see why Darwinists are so attached to the notion, and how experimental science has turned against them on this point. Then we will tie it all back to Sternberg's revolutionary way of conceptualizing the genome.

Neo-Darwinists generally concede that evolution's trial-and-error process of random mutations and natural selection is doomed to produce large amounts of waste material—junk DNA. And at one point these evolutionists seemed to have experimental confirmation. In the 1970s scientists discovered that only a small amount of the human genome codes for proteins, and many biologists concluded that the remaining DNA was mostly junk code that had accumulated over millions of years of evolutionary trial and error. Although some biologists warned against assuming this genetic material was useless junk, the idea spread rapidly through science journals and textbooks as important evidence of blind, trial-and-error evolution.

In 1976 Richard Dawkins summarized this view in his book *The Selfish Gene*. "The true 'purpose' of DNA is to survive, no more and no less," he wrote. "The simplest way to explain the surplus DNA is to suppose that it is a parasite, or at best a harmless but useless passenger, hitching a ride in the survival machines created by the other DNA."[15]

Year after year the view has been repeated by other researchers:

+ Lesley Orgel and Francis Crick, 1980: "Much DNA in higher organisms is little better than junk and can be compared to the spread of a not-too-harmful parasite within its host."[16]

+ Douglas Futuyma, 2005: "Only Darwinian evolution can explain why the genome is full of 'fossil' genes."[17]

+ Michael Shermer, 2006: "The human genome looks more and more like a mosaic of mutations, fragment copies, borrowed sequences, and discarded strings of DNA that were jerry-built over millions of years of evolution."[18]

+ Jerry Coyne, 2009: "We expect to find, in the genomes of many species, silenced, or 'dead,' genes: genes that once were useful but are no longer intact or expressed."[19]

+ John C. Avise, 2010: "Noncoding repetitive sequences—'junk DNA'—comprise the vast bulk (at least 50%, and, probably much more) of the human genome."[20]

Eventually this view trickled down to the general population, after geneticist Francis Collins plugged the idea of junk DNA in his popular 2006 book *The Language of God* to argue for evolution. By 2008, when I lectured in Switzerland about evolution's molecular-level problems, students from Fribourg University were in the audience, and a major objection from some of them, delivered with great confidence, was that junk DNA proved evolution—end of story.

But the students, and indeed even Francis Collins himself, were behind the times. In the first decade of the new century research findings had already rendered the idea of junk DNA untenable. Further progress in this decade has driven more nails into its coffin. At the moment, a race is on: Who can find the most function in the parts of DNA earlier deemed to be evolutionary junk?

Biologist Jonathan Wells's 2011 book *The Myth of Junk DNA* details the evidence that has piled up against the idea.[21] Here are some of the roles performed by DNA that was once considered to be non-coding junk:

+ They code for RNA that has an important regulatory role in the expression of protein-coding sequences.

+ Introns of eukaryotic cells code for small RNA molecules that participate in the work of protein-coding machinery by modifying chromatins—a complex of DNA, RNA, and proteins.

+ Some pseudogenes have turned out not to be pseudogenes. Instead, they regulate the expression of other genes.

+ Long repetitive DNA sequences, which make up about half of the human genome, have various functions in embryonic development, DNA transcription, synthesis of blood cells, and fat metabolism. They control gene expression in the alimentary track, mammary glands, and testicles. They also have an important role in placenta formation.

+ The length of a DNA sequence may affect its expression rate.

These findings also spell trouble for a favorite evolutionary argument for the common ancestry of humans and other mammals. Since DNA codes for function, it's expected on both a Darwinian and design framework that humans and chimps would have a lot of DNA in common. The two species are, after all, similar in many ways. But evolutionists have argued that there are stretches of junk DNA shared in common between, for instance, humans and mice. Why would a designer insert the same bit of junk code into a mouse genome and human genome? Surely he wouldn't. But it makes perfect sense, evolutionists argue, if the stretch of DNA occurred due to a copying error in a common ancestor, who then passed the stretch of junk code on to both its mice and human descendants. There is, however, a problem with the argument: Researchers are now uncovering many sophisticated functions for DNA previously deemed useless, including for the ancient repetitive elements (AREs) often cited as proof of common ancestry.[22]

According to a related argument, the common ancestry of apes and humans is evidenced by the fact that geneticists have identified a chromosomal fusion event in humans that reduced their total number of chromosomes from twenty-four (the same as apes) to twenty-three, a mutation that was apparently neutral—neither helpful nor harmful. But this discovery that two of our chromosomes became fused into one actually tells us very little. Think about it. If the mutation event had never occurred and humans currently had twenty-four chromosomes, just as apes do, our having the same number of chromosomes could be due to common ancestry, *or* it could be due to a designing intelligence pursuing a common design strategy for the two biological forms. Thus, if apes and humans presently had the same number of chromosomes, it wouldn't rule out common design any more than various cars each having four tires would rule out common design as the explanation for this common feature among cars. Humans might have diverged from apes and then experienced the chromosomal fusion event. Or humans might have been intelligently designed separately and then experienced the chromosomal fusion event, likely when their population was still small, making it easier for the mutation to spread through the entire population. Either scenario is logically possible.

The biological information found in humans has a lot in common with that found in apes, but a considerable amount of our genetic information is unique to humans. This is just what you would expect given that humans are distinct from apes while sharing many things in common with apes. This is true to a lesser degree of humans and mice, and to a still lesser degree with humans and flowers. We are all organic life forms. We have some things in common, and some qualities unique to a given species; and—what do you know—we have some biological information in common, and some that is distinct.

Moreover, the informational differences grow increasingly more pronounced on closer inspection. Ann Gauger, Ola Hössjer and Colin Reeves explain:

Scientists claim that our extreme genetic similarity with chimps (on the order of 98.7 percent identity) indicates we share common ancestry. This statement neglects several facts. First, our genetic differences are larger than that number represents. Common estimates of similarity are based on comparisons of the *single nucleotide changes only*, while other kinds of genetic differences are disregarded. In addition, noncoding regions of DNA—long thought to be nonfunctional "junk"—contain many kinds of genetic regulatory elements, some of which are species-specific. These species-specific regulatory elements make up a very small proportion of the total count of differences, but have a significant effect on how our genome works. For example, many of these regulatory elements are known to affect gene expression in the brain.[23]

A New Framework for the Genome

ABOUT FIFTEEN years ago Sternberg published two important papers together with University of Chicago biologist James Shapiro,[24] closely related to the issue of junk DNA. When Sternberg arrived in Helsinki in 2004, he was already a leading scientist on this issue. His two lectures at my campus, "Genomes as Complex Systems" and "Reorganizing the Genome: Information Generating or Information Shuffling" were excellent, state-of-the-art presentations. Sternberg summarized the revolution underway in the understanding of biological information systems with a slide representing a series of either/or contrasts that Shapiro had put forward in a recent journal article. As explained there, the twentieth-century understanding of genetics was an atomistic model, while the twenty-first-century understanding is a genome-centric model. The earlier framework was reductionist; the new framework, one of complex systems. The old model viewed biological operations as mechanical; the new model sees them as cybernetic. In the old model, the main focus of hereditary theory was "genes as units of inheritance and function." Now it's "genomes as interactive information systems." On the old model DNA was viewed as "a passive vehicle of genetic information" and an "active program during development." On the new model, DNA is viewed as a medium for storing data. On the old view, a common metaphor for genome organization is a string of beads. Within the new framework, it's a computer operating system.[25]

The point isn't that the genome is no more and no less than a computer operating system. Shapiro's point, I would surmise, is that biologists on the cutting edge have moved over to this new framework because it better illuminates some of the layers of architectural and informational complexity in the cell, layers obscured by the old framework. But the new framework, to be sure, has its own limitations. Computer operating systems are not sophisticated enough to make copies of their hardware and software—copies that can make copies, etc. They are not self-replicating. The cell is. But the new framework at least gets us much closer to the sophisticated reality of the genome and the cell. And here's what Sternberg came to realize: The new framework is a design-centric framework. Computer operating systems are, after all, designed.

9. Colleagues Dare to Explore

In 1985 I was sitting in a Mövenpick restaurant in Switzerland with my family and biochemistry professor Michael Gold of Portland, Oregon. In front of us was the house special, a huge ice-cream bowl, which we attacked from all sides. Later Gold wrote to me, "In the future when anyone tells me that the lignin biodegradation field is competitive, I will answer that they haven't seen the Leisola children eat together from a communal ice-cream bowl—internecine warfare."

Of all the scientists I have known over the years, Gold—who died in 2015—was one of the most talented and original. We first met at another meal, the breakfast table at a science meeting in Vancouver. I asked him whether he was the famous professor Gold. "Yes," he said, "and how is your sex-life?" I managed to keep my mouth from falling open and replied, "Not bad; I have four children." From that moment on we were friends. Once we were sitting on the beach in the city of Myrtle Beach, South Carolina. I asked him if he considered evolution an interesting theory or a fairy tale. He said that he had never really thought about it but that probably it was true.

I was surprised that an internationally known biochemist accepted evolution without ever having given it serious thought. But his case was not unusual. I have discussed evolution with dozens of colleagues in many parts of the world, and I find that very few of them are well-acquainted with even the basics of the theory. Most just accept it on faith.

Happily, Gold did not remain indifferent on the matter, and we had many good conversations about evolution in the years that followed.

Previous chapters relate various examples of evolutionists responding reflexively and dogmatically to my arguments against evolutionary theory, but I should emphasize that Gold was far from being the only pro-evolution colleague who proved willing to discuss evolution with me in an open and friendly manner. I would be remiss not to describe several such cases. They proved encouraging to me, and I share them in the hope that they will hearten others in academia who find themselves a lonely voice of dissent against a mostly close-minded Darwinian establishment. Be encouraged: There really are open-minded scientists out there.

In fact, I met another one at the same Vancouver meeting, organic chemist and University of Amsterdam professor Hans Schoemaker. With him I eventually published several papers and had many interesting discussions over the years about the origin of life—the most recent one in my sauna in Finland.

Some of the best conversations I had about evolution came during my time at the Swiss Federal Institute of Technology (ETH), where I arrived as a post-doctoral fellow in 1981. Professor Armin Fiechter (see Figure 9.1), the head of the Institute of Biotechnology and one of the pioneers in the field of biotechnology, invited me to lead a small research group, and later to give a seminar on the work I had done in Finland. In the seminar I briefly described my studies with bacterial plasmids and explained my views on the problems with the origin of life and the origin of biological information. Fiechter got upset and left the lecture room without a word, and he rebuked me the following day for discussing philosophy and not just facts. Actually, it would be more accurate to say that my talk had been too free of philosophy for his tastes. That is, I had not assumed philosophical materialism from the outset, and instead had simply let the facts and the evidence point up the failures of all existing materialistic explanations for biological information and the first life.

The good news is that later we became good friends, and he was a real support to me. My seminar planted a seed that eventually led to a

FIGURE 9.1—Professor and Mrs. Fiechter having supper with my children in 1985.

series of stimulating discussions among the scientists in the institute on the subject of evolution, which continued for over ten years, even after I had returned to Finland. Some years before his death in 2010, Fiechter invited me to write an article[1] about bioethics for a book series called *Advances in Biochemical Engineering*, where he was the volume editor and a member of the editorial board. "I know of no one else who could write it," he said.

Fiechter wasn't the only dyed-in-the-wool evolutionist at ETH who became a friendly interlocutor. Dr. Isaac Lorencez was the first of the institute's scientists who wanted to understand better what I had meant when I said in my seminar that "information is an essential feature of biology and there is no evolutionary explanation for its origin." Steeped in a world of evolutionary faith, he first found it difficult to take my view seriously, but after getting rid of the old ballast he quickly grasped the information-theoretical and mathematical problems and turned his attention to mathematical modeling, trying to understand biological organisms from the standpoint of information.

One of Lorencez's friends, a molecular biologist, wanted to meet me to convince me of the truth of evolution. After discussing the nature of genetic information, he admitted that he had never met anyone who looked at the genome from such a perspective. Dr. Duane Ulmer, an American post-doctoral fellow in my group, followed the many discussions I had with my fellow faculty members and admitted he was sur-

FIGURE 9.2—The evolution skeptics at ETH-Zürich: Roland Waldner (left), Andreas Muheim (right), and me (center).

prised the molecular biologists did not seem to have sound arguments for their Darwinian views. Another colleague in my seminar told me it had opened a new dimension in his thinking.

Five students pursued their doctoral dissertations as part of my research group, and the mechanisms of evolution were a repeated theme of coffee-table discussions. One of them admitted, after reading Michael Denton's *Evolution: A Theory in Crisis*,[2] that he now understood the problems with the theory. Two others, Andreas Muheim and Roland Waldner (see Figure 9.2), became critics of Darwin's theory.

While I lived in Switzerland, many internationally known scientists visited my home and, without exception, evolutionary mechanisms were the topic of discussion. At ETH I got acquainted with evolutionary studies of viruses. One colleague was developing a two-stage biological reactor to study the evolution of the $Q\beta$-virus of the *E. coli* bacterium under various selection pressures. In the first reactor *E. coli* was cultivated. In the second reactor it was contaminated with the virus. Both reactors were operated as continuous cultures. At one point, the experimenters increased the flow rate. *Flow rate* refers to the speed at which nutrients are pumped into the reactor and the produced biomass is pumped out. The volume remains constant. Increasing the flow rate forces the organisms to grow faster, up to certain natural limits. When the flow rate was increased, both bacteria and viruses had to multiply faster, but here is the telling part: With the increased flow rate the virus produced only those parts of its structure it absolutely needed for survival. That is, it tended to shed biological information, not generate new biological information. The results reinforced an earlier study, from 1967, which found that the virus lost 83% of its size when forced to multiply rapidly.[3]

Studies carried out with Qβ-virus are sometimes heralded in biology textbooks as examples of evolution in action. What gets obscured in such triumphant announcements is this: No new biological information was formed during the experiments, and much was lost. I asked my colleague how his experiments, even in principle, demonstrated that the evolutionary mechanism could generate new form and information. He admitted he did not see how it could be interpreted in this way. How can loss of biological information be interpreted as gain of information?

The conversation was an instance of something I ran into frequently: scientists willing to have frank, open-minded conversations with me about evolutionary theory, but only in private. I came to understand through my many international connections that neo-Darwinism, while little valued among mainstream biologists who spent any time thinking about the theory, was treated by them as a third rail—too dangerous to touch. Many who understand one or more of the problems with it are afraid to share their views for fear of losing their positions.

Evolution Experiments with Bacteria

I WAS once sitting in the campus restaurant at ETH with Dr. Lorencez discussing various evolutionary hypotheses when Dr. Branko Kozulić, a new scientist at the Institute, joined us. When I explained to him the topic of our discussion, he immediately gave a speech against evolution. He said that he likes to surprise his scientist colleagues with such views.

FIGURE 9.3—Croatian biochemist and inventor Dr. Branko Kozulić, my colleague and friend from ETH-Zürich days.

Kozulić is a talented and versatile biochemist who has over fifty patents. He later worked at private biotech companies. Kozulić's major arguments against evolutionary theory are related to the huge complexity of biological metabolic informational networks and their regulation.

Kozulić analyzed the literature on sequenced genomes and concluded that each species has hundreds of what are termed ORFan genes or singleton genes. These are genes with no resemblance to those found in other taxa (categories of organisms such as species, genera, and families).[4] Each such gene is a huge challenge to the theory of evolution. Evolution's gradually diverging tree of life predicts that genes in one taxon (singular of taxa) will typically have "sister" and "cousin" genes in closely related taxa, genes that are quite similar, with the differences only growing pronounced as you move further apart on the tree of life to quite different plant and animal forms. This goes back to the idea that evolution progresses by a series of small random mutations to DNA. ORFan genes contradict this evolutionary prediction.

Could blind evolution possibly make a great leap from one gene to a very different ORFan gene, and so eliminate the need for a series of small random mutations and an extended series of intermediates? In 2015 Kozulić and I made a careful analysis of studies by Nobel laureate Jack Szostak's group, and we concluded that even with extremely generous assumptions the probability of a random process landing on functional activities among random RNA or protein sequences is so low that it represents a practical impossibility.[5]

That is all background to what happened next. One time a scientist from New Zealand joined our discussion. He was upset by our views and told us he knew of experiments that proved us wrong. Specifically, he referred to studies carried out by biologist Robert Mortlock[6] in the 1960s. Mortlock had shown how mutated bacteria learn to use rare sugars such as D-arabinose and xylitol as a food supply.[7] But Mortlock's observations, while interesting, only revealed natural variability among bacteria. The phenomenon is never the result of new genetic information. (See the discussion in Chapter 3 above on xylitol-eating mutant bacteria.) Instead, it appears to be the result of damage to normal control systems and to promiscuous enzyme activity. ("Promiscuous enzyme activity" refers to the pre-existing ability of many enzymes to catalyze a reaction associated with another substrate, but at a much reduced rate.)

All that being said, if we are going to try to observe macroevolution in action, and not just retell imaginative stories about what it must have done in the distant past based on naturalistic assumptions, the bacterial world is a great place to go. Bacteria are real mutation machines. Under ideal conditions some can multiply in about ten minutes, so they are ideal organisms for the study of evolutionary mechanisms.

Biologist Richard Lenski has been doing that in his lab at Michigan State University for decades now. His group's experiments with *E. coli* are probably the best-known long-term evolution-simulating experiments. By now they have cultivated more than 68,000 generations, an impressive achievement. If you trust some reports, their long-running experiment has already decisively shown the powers of unguided evolution. "A major evolutionary innovation has unfurled right in front of researchers' eyes," the *New Scientist* reported in 2008. "It's the first time evolution has been caught in the act of making such a rare and complex new trait."[8]

Before we delve into that claim, notice that for 150 years we have repeatedly been told that the grand powers of the mutation/selection mechanism have been proven beyond a shadow of a doubt. And yet here, in 2008, a prominent science journal reports that a lab has uncovered *the first evidence* of evolution's ability to innovate in an impressive way. The implication shouldn't be missed: All the grand claims for evolution that came before this lacked empirical support.

Then, of course, we need to ask: Is this new claim the real deal, or more bluffing? Biochemist Michael Behe offers this assessment:

> Nothing fundamentally new has been produced. No new protein-protein interactions, no new molecular machines…. some large evolutionary advantages have been conferred by breaking things. Several populations of bacteria lost their ability to repair DNA. One of the most beneficial mutations, seen repeatedly in separate cultures, was the bacterium's loss of the ability to make a sugar called ribose, which is a component of RNA. Another was a change in a regulatory gene called *spoT*, which affected en masse how fifty-nine other genes work, either increasing or decreasing their activity. One likely explanation of the net good effect of this very blunt mutation is that it turned off the energetically costly genes that make the bacterial flagellum, saving the cell some energy. Breaking some genes and turning others off, however, won't make much of anything.[9]

Put simply, for bacteria, this is "evolution" by losing or damaging genes.[10] Members of the species compete to grow and reproduce the fastest, and tend to shed functions that aren't immediately useful.

Diving Deeper

LET'S TAKE this dive into the microbial realm a bit deeper now. We'll try to make it as accessible as possible, but if you're a non-scientist and find it heavy going, feel free to skip down to the chapter's final subheading, or even to the final paragraph and the chapter's take-home bullet points there.

One interesting development in the Lenski experiment was reported in 2008. Normally, *E. coli* cannot use citrate as a nutrient when oxygen is present. But some of Lenski's *E. coli* gained the ability to use citrate when oxygen was present. Intriguing; but we have to ask: How, exactly, did it gain this ability? The lab eventually determined that a gene, *CitT*, which encodes a protein that normally imports citrate into the cell only when oxygen is *not* present, had mutated. The mutation gave the protein the ability to import citrate even when oxygen was present. Notice that the protein already had the ability to import citrate. Behe, again, ably puts the experimental results in perspective: "It was an interesting, if modest, result—a gene had been turned on under conditions where it was normally turned off."[11] Casey Luskin elaborates: "What really happened? A switch that normally represses expression of *CitT* under oxic conditions was broken, so the citrate-uptake pathway got turned on. This isn't the evolution of a new molecular feature. It's the breaking of a molecular feature—a repressor switch."[12] So again, a minor innovation was achieved by breaking stuff. This is no way to build a cathedral. Or a new animal. Or a new plant. Or a new kind of cell. Or even a novel protein.

Chapter 3 described results from a 1965 experiment that showed that the *Aerobacter aerogenes* bacterium had learned to grow on xylitol. This was billed as an example of evolution's great powers. What had actually occurred in the experiment? One mutation had destroyed the normal regulatory system in the bacteria, resulting in a continuous production of one of its enzymes capable of oxidizing xylitol. There, as in the *E. coli* experiment, nothing new was created.

And again, this isn't for lack of trying on the part of the bacteria. Bacteria grow rapidly. Some thermophilic bacteria can multiply in ten minutes, and random mutations in the genome are directly transferred to the next generation. The mutation rate is easy to increase by ultraviolet light or toxic chemicals. This makes bacteria an ideal tool to explore the powers of evolution. But as Alan Linton, a professor of bacteriol-

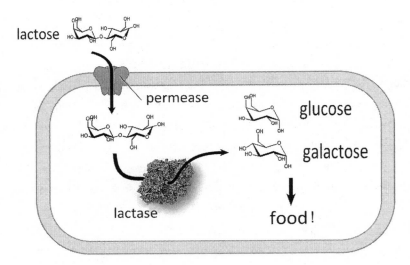

FIGURE 9.4—Permease transports lactose into *E. coli* where lactase degrades it into glucose and galactose.

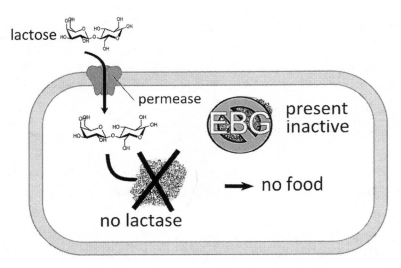

FIGURE 9.5—When the lactase gene is inactivated, the bacterium cannot use lactose as food.

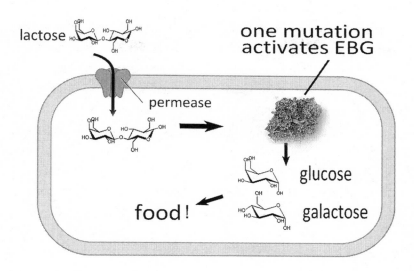

FIGURE 9.6—One mutation in the *ebg*-gene recovers the ability to use lactose.

ogy from Bristol University, writes, "Throughout 150 years of the science of bacteriology, there is no evidence that one species of bacteria has changed into another."[13]

E. coli experiments carried out by Barry Hall[14] shed more light on this. The bacterium can digest the milk sugar lactose for an energy source. To do this it has a permease on its cell membrane to transport lactose into the cell and an enzyme called lactase (beta-galactosidase) to split lactose into two simple sugars, glucose and galactose. (See Figures 9.4–9.6.) Hall destroyed the lactase-encoding gene, which resulted in mutants that could no longer use lactose as the energy source. When Hall cultivated these mutated cells in lactose-containing nutrient solution, mutants regularly appeared that could grow on lactose. What had happened? *E. coli* has an enzyme (Ebg) that closely resembles the lactase enzyme, though it is normally not able to split lactose. A single mutation in *ebgA* (the gene that codes for Ebg) is sufficient to allow slow growth on lactose. This mutation, while useful in this narrow situation, represents the tiniest of steps: A single point mutation to Ebg, the role of which is unclear, allowed it use lactose, although according to Hall "the best

Ebg enzyme does not even approach the catalytic efficiency of the LacZ enzyme" for digesting lactose. Such results are neither controversial nor impressive: A single point mutation in a large bacterial population with billions of cells is well within the reach of random evolution.

The most impressive evolutionary experiment to my knowledge so far reported was carried out by an international team using *Salmonella enterica*.[15] On October 22, 2012 a report claimed that this was the *first time* a group demonstrated the origin of a new gene.[16] In reality a gene with a weak side-activity was duplicated and the side-activity was strengthened. Intriguing, but nothing more—and nothing new. Yet what follows is how the work was described in the popular press (emphasis added to show where intelligent engineering was introduced into the experimental environment):

> Researchers *engineered* a gene that governed the synthesis of the amino acid histidine, and also made some minor contributions to synthesizing another amino acid, tryptophan. *They then placed* multiple copies of the gene in Salmonella bacteria that did not have the normal gene for creating tryptophan. The Salmonella kept copying the beneficial effects of the gene making tryptophan and over the course of 3,000 generations, the two functions diverged into two entirely different genes, marking the first time that researchers have directly observed the creation of an entirely new gene in a *controlled laboratory setting*.[17]

There has been another interesting evolution experiment carried out using *E. coli*. The theoretical background to the experiment is as follows. It is generally assumed that a multi-step mutational evolutionary path is possible if all the intermediary steps are functional and can each be reached by a single mutation. The activity produced in this way may, however, be so weak that the cell must over-express the hypothetical newly formed enzyme—in other words, produce too much of the enzyme, causing a huge strain on the cell because it has to use extra synthetic capacity for this. Therefore it is likely that the cell would shed such a weak side-activity. The modest benefit wouldn't be worth the strain caused by the overproduction.

Ann Gauger and her colleagues[18] studied what happened in such a case under laboratory conditions. They introduced a mutation that partially interfered with a bacterial cell's gene for the synthesis of the amino acid tryptophan. Then they introduced a second mutation into the gene that completely abolished the ability to synthesize tryptophan. Cells with the double mutant could, theoretically, regain weak tryptophan-synthesizing ability with only one back-mutation. Given more time, cells with the one back-mutation might then undergo one more back-mutation to regain full tryptophan-synthesizing ability. This might demonstrate how a cell could gain a new function with just two mutations. But this did not happen. Instead, cells consistently acquired mutations that reduced expression of the doubly mutated gene. The experiment suggests that even if the cell could acquire a weak new activity by gene mutation, it would get rid of it because weakly performing functions of this sort exact too heavy an energy burden.

So, while the described experiments are often promoted as evidence for neo-Darwinian evolution, they either (a) are intelligently designed and do not accurately reflect what happens in nature, or (b) underscore the narrow limits of neo-Darwinian evolutionary change.

Had Cells but World Enough and Time

RECENTLY AN important paper was published by Chatterjee et al. evaluating the time-scale needed for evolutionary innovations. They found that starting from a flat point in the fitness landscape (exactly the situation that exists when starting from a random sequence), the search for a functional sequence does not succeed for a typical sequence of 1,000 nucleotides, even if multiple populations could search throughout the entire age of the earth. The search is unsuccessful even when the target area is very broad: "The estimated number of bacterial cells on earth is about 10^{30}. To give a specific example let us assume that there are 10^{24} independent searches, each with population size $N \sim 10^6$. The probability that at least one of those independent searches succeeds within 10^{14} generations for sequence length L=1000 and broad peak of c =1/2 is less than 10^{-26}."[19]

The search also proves unsuccessful when the targets are many and broad, regardless of what population size is modeled. The study thus supports the conclusion that there is a strict limit to what unguided evolutionary processes can achieve.

The meaning of modest bacterial mutations is repeatedly oversold in the world of naturalistic evolutionary faith, much as Stanley Miller's origin-of-life experiment was oversold. It calls to mind an old saying in Texas—all hat and no cattle. In other words, all talk but no real results.

Decades of evolution experiments with microorganisms can be summarized as follows:

+ Microorganisms can be mutated to overproduce desired compounds.

+ Production of such organisms requires huge mutation rates along with systematic artificial selection or tedious construction and loading of efficient pathways into organisms, neither of which occurs in nature.

+ Such organisms are not viable in natural conditions due to high mutational load or extra non-natural genetic load.

+ Isolated microbial populations in laboratory experiments vary within narrow limits and lose information over time.

I know several successful academic biologists willing to concede all this, and to puzzle over all of it in stimulating conversations in the hallways of international conferences. But very few of them are willing to do so in public. The enforcers of Darwinian orthodoxy still have the power to threaten careers and, in some situations, to deliver on those threats. This is how the guardians of the old orthodoxy defend the citadel—not with fresh evidence but with fear.

10. MECHANISMS
MALFUNCTION

I N 1992 THE MANAGEMENT TEAM OF BIOTECH COMPANY CULTOR PRE-
sented a challenge to its research center to manufacture the tooth-
friendly sugar alcohol xylitol directly from glucose. My first reaction was
that the task would be extremely difficult at best. I knew that some Can-
dida yeasts can reduce the five-carbon xylose sugar to xylitol via a one-
step enzymatic reaction, and I had with my team improved such yeasts
by random mutations to overproduce xylitol.[1] But there were no organisms that could convert the six-carbon sugar glucose—a cheap sugar—to
xylitol. Nonetheless we provisionally accepted the challenge.

One of my first steps was to study the metabolic pathway map[2] with
my research team to figure out if there were theoretical possibilities to
direct the flow of carbon from the six-carbon glucose to a five-carbon
xylitol. The central metabolism of yeasts has a link between glycolysis
and the pentose phosphate cycle to produce the five-carbon ribose and
deoxyribose sugars needed for DNA and RNA. That meant that we
might be able to use this knowledge and overproduce five-carbon sugars
from glucose.

A literature search revealed that there are yeasts that accumulate
arabitol in certain conditions. This was promising because arabitol is
closely related to xylitol. (See Figure 3.4 in Chapter 3.) So my research
team developed a plan that involved cloning two enzymes (D-arabitol
dehydrogenase, which forms D-xylulose from arabitol, and xylitol dehydrogenase, which forms xylitol from D-xylulose) into a yeast strain

that can tolerate high sugar concentrations. For that we needed a suitable yeast strain, the necessary genes, and genetic engineering techniques to transfer the genes into a chosen yeast. We would also need to carefully consider the redox balances, reaction equilibria, and cultivation conditions.

Happily, after six months of work on the problem, my team had demonstrated that the chosen yeast can produce xylitol directly from glucose, an invention for which we eventually received a patent.[3]

Our invention of the new yeast was just one of many experiences in which I saw firsthand the crucial role of foresight, planning, and expert design in bio-engineering a new organic form from existing ones. My research has often focused on modifying proteins—mainly enzymes—to function better in specific industrial processes, and on altering the metabolism of microbes so they would produce various chemicals economically. More than forty years of work in this field have left me more skeptical than ever of theories of blind, unguided evolution. It is increasingly clear to me that random mutations cannot produce novel functional information—even one new gene—with or without help from natural selection, and with or without help from any of the other ancillary mechanisms proposed to rescue neo-Darwinian evolutionary theory from the swelling onslaught of contrary evidence. (More on those at the end of this chapter.)

One source of that contrary evidence comes from advances in genetic engineering. New tools have made it possible to quickly elucidate the genetic program of an organism. One of the tools for sequencing genomes involves restriction enzymes, which make it possible to cut DNA at specific sites. Another, which won its inventor a Nobel Prize, allows for rapidly copying a DNA sequence many times. It's known as the polymerase chain reaction (PCR), and I still vividly remember first learning about it at a private seminar[4] in Japan where it was introduced by the editor-in-chief of *Nature Biotechnology*. Since that time, this and other genetic tools have been employed to help criminal investigations, discover heritable diseases, and track kinships and the historical move-ments of groups of people.

More specifically for our concerns here, protein engineering tech-niques help researchers modify proteins and microorganisms and test how much random mutations can change genes and organisms. In this way the techniques promise to shed new light on the powers, and limits, of the neo-Darwinian mechanism.

In lab-directed evolution using classical mutation techniques and re-cently developed genetic tools, random changes are made in a gene. Then out of the thousands or even millions of resulting mutants, researchers probe for and select better variants for a given purpose. The properties of many enzymes and other kinds of proteins have been modified in this way. The experimental results suggest that an enzyme or other protein can be modified in limited ways without destroying its structure, but one basic structure has never successfully been changed to another basic structure, even though molecular biologists avail themselves of some-thing the neo-Darwinian mechanism lacks: the power of foresight, plan-ning, and design.

I explained these observations in Japan at the Rare Sugar Congress in 2006, and I wrote a review article on the limits and possibilities of protein engineering in 2007.[5] (See Figure 10.1.) Six years later I was in on a Ph.D. examination committee at Ghent University in Belgium. The topic of the doctoral thesis was modification of two enzymes so that they could function as tagatose epimerase (tagatose is a low-calorie sweetener).[6] This would allow tagatose production from readily available cheap sugars like glucose or fructose. The candidate used in his studies two distinct approaches: random modification of the respective genes, and systematic design. He also combined the two approaches. None of his efforts produced the desired changes even though several million mutants were tested.

It's a topic for another place why the systematic design and blended approaches failed in his experiment. But I saw in the failure of the ran-dom approach a teachable moment—or maybe I was just feeling mis-

FIGURE 10.1—Rare Sugar Congress in 2006, held in Takamatsu, Kagawa, Japan. Pictured are the members of the society's international committee. Left: Dr. Erick Vandamme from Ghent University in Belgium; Dr. Masaaki Tokuda from Kagawa University, Japan; origin-of-life researcher Dr. Arthur Weber from NASA Ames Research Center; and Dr. Saisamorn Lumyong from Chiang Mai University, Thailand. Right: President of the International Society of Rare Sugars, Dr. Ken Izumori from Kagawa University, Japan; Matti Leisola; and Dr. Deok-Kun Oh from Konkuk University, Korea.

FIGURE 10.2—Examination committee members in February 2013: Dr. Jo Maertens (committee member), Dr. Wim Soetaert (dissertation director), Em. Prof. Dr. Matti Leisola, (committee member), Em. Prof. Dr. Erick Vandamme (committee member), and Koen Beerens (doctoral candidate). [Prof. Dr. Chris Stevens (committee chairman), Prof. Dr. Els Van Damme (committee secretary), and Prof. Tom Desmet (assistant dissertation director) are not shown in the figure.]

chievous; in any case, I asked the candidate whether he had calculated the probability for the desired changes in a case where he needed only two or three specific amino acid changes. Since he had not done the calculation, I told him. The probability for only two simultaneous specific mutations in the enzymes he was studying was not better than one chance in twenty million, and that was a generous, simplified estimate. Given various additional factors that a more nuanced calculation would take account of, he probably would have needed many more attempts for the desired purpose. So it was no surprise he could not find a positive result with the random method. I was later told that the university had hesitated to invite me to the committee upon learning what I think of the power of the neo-Darwinian mechanism. During the examination, however, some of the members of the committee referred to my probability calculations and admitted the weaknesses of the random approach.

About the same time, an article was posted on a Ghent University webpage[7] referring to a paper published in the journal *PLOS/Biology*.[8] The article posted on the webpage insisted that the problem of new innovations had now been solved and that criticism of the neo-Darwinian mechanism had been proven wrong. Was the timing a coincidence, or was the university underscoring its Darwinian purity in the wake of the student's research and my comments as part of the committee, which together suggested an embarrassing shortcoming in the creative powers of the neo-Darwinian mechanism? Without a public relations counteroffensive, it might have appeared that Ghent University was endorsing the heterodox practice of Darwin-doubting. Many universities will go to great lengths to avoid any such appearance of impropriety. Why? Richard T. Halvorson explained it this way in the *Harvard Crimson*:

> Intellectual honesty requires rationally examining our fundamental premises—yet, expressing hesitation about Darwin is considered irretrievable intellectual suicide, the unthinkable doubt, the unpardonable sin of academia.
>
> Although the post-modern era questions everything else—the possibility of knowledge, basic morality and reality itself—critical discussion of Darwin is taboo.[9]

Whatever the reason for the timing of the Ghent University article, the article itself worked very hard to assure readers that all was now right in the land of Darwinism. It asserted that "an important unsolved question in Darwin's theory of evolution is how new properties seem to appear from nowhere," but that this problem had finally been solved in a study published the previous December about the evolutionary history of certain enzymes.

However, enzymes are one of my areas of specialization, and when I examined the study firsthand, I found the research results were actually quite modest. "The preduplication ancMals enzyme was multifunctional and already contained the different activities found in the postduplication enzymes, albeit at a lower level," the researchers reported. In other words, the results confirmed what was already known: variation (microevolution) can modify the relationships of side-activities already present in an enzyme. The research had thus not shown the origin of new activities and the result was in no way new.

As for the article on the University of Ghent webpage, it was removed after molecular biologist Douglas Axe commented on it in an essay appropriately titled "Belgian Waffle."[10]

Something Rotten in the State of Rot Science

As AN enzyme scientist and Darwin skeptic, I keep my eyes peeled for articles making ambitious claims for enzyme evolution. A case in point: Helsinki University ran a news piece announcing, "A fungus closed the oil taps: A study published in *Science* strengthens the view that fungi had a role in stopping the growth of coal deposits."[11] The news piece went on to summarize the journal article's central argument, namely that a certain enzyme's evolutionary history helps explain why carbon was laid down in the geological column quite generously in an earlier period but no longer. Curious, I went to read the journal article itself, which details a major research study conducted by seventy scientists and published in 2012. The paper suggests "that the sharp decline in the rate of organic carbon burial at the end of the Permo-Carboniferous was caused, at least in part, by the evolution of lignin decay capabilities in white rot Agaricomycetes."[12] In other words, thanks to the powers of evolution, the white rot Agaricomycetes got so good at rotting wood—due specifically to the evolution of oxidative enzymes—that wood no longer stood much of a chance of surviving long enough to become coal.

But comparison of the variation in, and number of, similar enzyme activities does not tell how these activities first arose. In this major comparative study, the origin of oxidative enzymes remained unexplained. Also, even rot fungi bubbling with oxidative enzymes cannot degrade lignin in the absence of an energy source, as we showed in our review article, "Lignin: Designed Randomness."[13] Without oxygen there is virtually no decay.

This brings us to another and better explanation for the coal deposits: Wood was buried under oxygen-free earth layers. Since lignin degradation is a highly oxidative process, the lack of oxygen gave nature time to convert the wood to coal.

So the study provided no adequate cause for the evolution of the enzymes in question, and an adequate cause for the carbon-rich layers in the Earth's crust is already known. But tell a good evolutionary just-so story festooned with some rigorous-sounding references to molecular clock calculations, and the Darwinist cheering squad is sure to take the field. Yet at some point the pompoms and megaphones and cheering cease to impress.

A Microorganism is to a Maserati as a…

THE ARTICLE did not describe a credible way for oxidative enzymes to have evolved, but is there a case to be made for it? No, and for the same reasons I am skeptical of macroevolution at the microbial level generally. For one thing, Axe has made a thorough analysis on the origin of novel protein structures and shown, against the conventional view, that not even a single new protein structure could be formed by a blind evolutionary search.[14] More on that and other protein evolution research below. Here, suffice it to say that Axe's finding spells trouble for neo-Darwinian evolution at every stage of the journey from the first cell to the cacophony of life forms we find on planet earth today.

What about changing a microorganism to a different one, an even taller order than creating a new protein? Microbes are good targets for the study of evolutionary claims because they can be freely mutated with various methods and, as we discussed earlier, they reproduce rapidly in numerically large populations, leaving lots of room for random mutations to run wild and possibly stumble onto a useful innovation. So what has all this shown us? Random mutations have been used in the history of biotechnology to produce numerous strains, and from these have been produced large amounts of various compounds such as organic acids, enzymes, and vitamins. But as impressive as all this is, the gains are usually made via injuries to control mechanisms. That is, such niche specialists are created by breaking or compromising certain parts of a genome. Such organisms are usually not viable in nature. What we don't see is the generation of novel structures and novel biological information—a prerequisite for an evolutionary mechanism able to generate the diversity of life we find around us.

Genetic technologies have made possible the addition of new capabilities to some types of organism. Baker's yeast has been converted to produce xylitol, and lactic acid bacteria to make non-natural rare sugars. Vitamin C is produced by genetically modified bacteria, and other engineered bacteria can produce indigo blue. But a prerequisite for the function of all these intelligently designed industrial organisms is that their basic metabolism is not disturbed. A living cell resists change beyond certain modest limits because it is an extremely complicated and carefully controlled holistic entity.

Consider a sports car. You could turn loose a kid with a can of lime green spray paint on the car's beautiful factory paint job, and the car would turn more heads as you roll down the boulevard. You could rip out the passenger seats to give the car more storage room and improve acceleration and fuel efficiency. You could yank out the radio to create an extra coin holder. Make any of these changes, and the car would still burn up the streets on a Saturday night. But start making bold, random changes to the engine design or transmission, and the car will excel only at gathering dust. That's because the engine and transmission are key parts of the machine's complex and interdependent design. To redesign either requires skill, foresight, and planning—intelligent design. A microorganism is the same way. You can tinker around the margins, but start making significant random changes, and you have a problem.

In fact, this illustration really undersells the challenge. A microorganism is to a sports car as a sports car is to a garden spade—vastly more sophisticated. No wonder Evelyn Fox Keller, in her Harvard University Press work *The Century of the Gene*, wonders, "How can a process dependent solely on the chance appearance of new mutations have given rise to structures whose function is to provide pockets of resistance to the disordering forces of chance—structures designed, that is, to be robust?"[15]

BIO-Complexity

As I mentioned earlier, in 2008 Douglas Axe contacted me about starting a new science journal. I pondered the suggestion for quite a while and finally agreed to be the first editor-in-chief. My condition was that all the published material would be scientific and carefully peer-reviewed, and that the journal would not get involved in the political-religious-atheistic discussion running wild around the topic of biological evolution. This was no sticking point, as it turned out, because it was right in step with Axe's hopes for the journal. We decided to invite to the editorial board all the possible interest groups and sent invitations to a total of 127 evolutionists, intelligent design theorists, and persons neutral on the topic.

FIGURE 10.3—Dr. Douglas Axe and me.

The journal was christened *BIO-Complexity*. It is an open-access journal, and the articles can be read at bio-complexity.org. We did manage to get some evolutionists to agree to serve as peer reviewers, but unfortunately, not a single proponent of blind evolution accepted the invitation to join the editorial board, although many were polite and wished us success. I knew several of these scientists professionally, and I suspect that at least some of them would have liked to be part of the enterprise but understood the heavy price they might pay from the enforcers of evolutionary orthodoxy if they joined the board.

The journal never aimed at large numbers of articles, but at publishing thoroughly peer-reviewed, high-quality, original research and reviews. The goal was to send every manuscript accepted for peer review to a scientist critical of intelligent design. Every published article is also open to online comment. During my five years as the editor, sixteen articles were published and fourteen rejected. Not a single article was published without peer review and significant modifications.

Two examples of the peer-review process are instructive. An evolutionist reviewer commented on the paper by Gauger et al.[16] as follows: "I think that this article is very important and should be published in a high impact journal like PNAS so that evolutionary biologists cannot overlook it."

My own review article on lignin[17] received positive reviewer feedback, but one reviewer commented:

> The crux of the article is to present an argument for intelligent design—a concept that I believe is logically refutable. Accordingly, I have decided it would be hypocritical of me to provide a careful review of the article. Very briefly though, the review of what has been scientifically determined regarding lignin biosynthesis and decay was clearly written and enjoyable to read. Regrettably, I find logical and philosophical inconsistencies in other aspects of the paper.

I wish the reviewer had told me how, according to him, my conclusions were self-contradictory. That would have made the review both useful and scientific. Unfortunately, he did not do so.

He was at least civil. Some of those responding to the published article were practically frothing at the mouth. I responded calmly to the charges,[18] but it wasn't easy. Some asserted that even a superficial review of the relevant scientific literature would prove our claims wrong. But the conclusions were based on my thirty years of research and publication in this specialized area. I knew the literature more than superficially. We asked the critics to name something in the literature that proved us wrong, and we got only more bluffing and frothing. Such behavior speaks to the fury of the naturalistic internet police against everything that hints of design. I take comfort in the fact that we are not alone in such matters. Philosopher Thomas Nagel and biologist James Shapiro have experienced the same rage when they have had the courage to criticize neo-Darwinism, although both consider evolution true and Nagel says he is an atheist.

How to Modify a Protein

LET'S DELVE a little deeper into the question of protein evolution, since proteins, including enzymes, are foundational to life. Proteins are natural polymers made of amino acids. (See Figure 10.4.) Nature, recall, uses twenty different amino acids to build the protein main chain. We can think of this informally as a twenty-letter alphabet. The bond between the amino acids is always the same—a peptide bond. The side chains of amino acids vary. Just as words and sentences are built from various letters, different protein chains are built from the twenty different amino acid letters.

The average protein is about 300 amino acids in length—more precisely, 267 for bacterial and 361 for eukaryotic proteins.[19] These chains of amino acids can be ordered in 20^{300} different ways, a figure we can also represent as 10^{390}. That's a 1 followed by 390 zeroes. Pause for a moment to grasp how big that number is. A single water droplet of average size contains some five sextillion atoms (5.01×10^{21}). There are an estimated 10^{82} atoms in the visible universe, a universe containing more than 100 billion galaxies; and galaxies have, on average, about 100 billion stars.

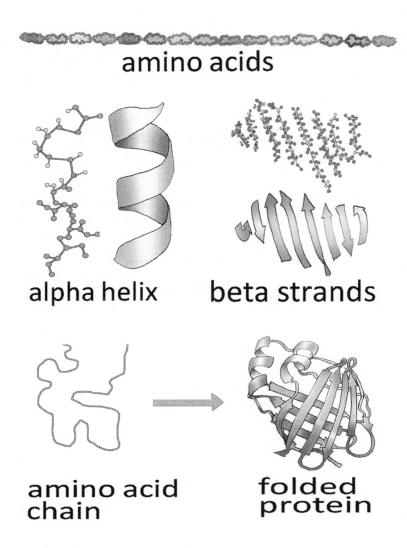

amino acids

alpha helix   beta strands

amino acid
chain

folded
protein

FIGURE 10.4—Protein chains vary in length, but are, on average, about 300 amino acids long. Proteins have some regular structures like alpha-helixes and beta-strands. In the cell the protein is folded into its 3-D functional form either spontaneously or with help from chaperones.

And yet the even number of atoms in the entire visible universe is utterly dwarfed by the possible letter combinations in an amino acid chain 300 units long.

Most of these possible sequences would yield dysfunctional rubbish. How many would yield a functional protein? Scientists generally agree, based on several lines of experimental data, that more than one specific protein sequence is capable of performing a particular function. But is there one functional sequence in ten? One in a thousand? One in a million? How rare are functional sequences in the larger sea of all possible sequences for a chain of amino acids? The question is relevant because if evolution is supposed to have stumbled on new functional sequences at random, we need to know how hard the lottery is, so to speak, in order to assess the plausibility of the chance hypothesis.

Scientists still debate the size of the fraction of functional protein molecules among all the possible non-functional sequences, as well as how best to describe the functional information residing in proteins. The difficulty is confounded by experimental findings showing that there are protein families with over 100,000 members having related but different sequences and, most likely, essentially the same structure and function. Moreover, some proteins have different structures but similar functions.

Laboratory investigations have returned varying figures, but the general drift is that functional sequences make up an almost unimaginably small fraction of the total possible sequences. The estimate of Taylor et al. is that a library of 10^{24} members should contain an AroQ-mutase enzyme.[20] Other studies yielded even longer odds of finding the proverbial needle (that is, a functional sequence) in the haystack of all possible sequences. Hubert Yockey,[21] based on reported cytochrome c sequences, estimated that this fraction is 1 in 10^{65}. Thirteen years later John Reidhaar-Olson and Robert Sauer[22] estimated that the fraction is 1 in 10^{63}. Fourteen years after that, Axe[23] concluded from his studies with penicillin-degrading β-lactamases that the probability of finding a functional enzyme among random sequences may be as low as 1 chance

FIGURE 10.5—Stepwise mutation of alpha-spectrin SH3 protein domain toward the B1 domain of streptococcal protein G does not succeed because the sequence space between the two is enormous and crowded with dysfunctional intermediates.

in 10^{77}. And eight years after that, in a study of four large protein families, Durston and Chiu[24] estimated that functional sequences occupy an extremely small fraction of sequence space, in all cases lower than 1 in 10^{100}. Long odds indeed. To return to our earlier illustration, the odds of randomly selecting the winning atom from among all the atoms in the entire visible universe is one chance in about 10^{82}, horrendous odds but still much better than 1 in 10^{100}.

The standard Darwinian response is that evolution starts with small polypeptides (which they argue could conceivably form by chance), polypeptides that have some function, and then a combination of duplication, mutation, and selection builds larger functional polypeptides. This is, however, pure speculation and the following paragraphs help to explain why this proposal doesn't work.

We can learn a lot about the chances of a random evolutionary search by investigating results from efforts to modify enzymes for industrial purposes. Enzymes are used in a host of applications—washing powders, food manufacturing, textile production, animal feed, and chemical production, to name only a few. Natural enzymes are not always suitable for industrial conditions where high temperatures, extremes of pH, and various chemicals interfere with enzymatic reactions. But by using the tools of genetic engineering it is possible to modify existing enzymes, tailoring them to specific industrial situations. The various methods fall into two large categories: random modifications and designed modifications. Let's take each in turn and consider what light they throw on the question of unguided evolution.

Random modification of existing enzyme structures. One approach is to randomly mutate the gene coding for a given enzyme. The process is called directed evolution, a confusing label since chance plays a leading role in this approach, and because the term borders on being an oxymoron: after all, the term *evolution*—at least in its contemporary scientific context—describes an undirected or blind process of change. The approach is *directed evolution* in the sense that the scientists who are re-engineering the enzyme decide which enzyme type to start with and in what direction to drive it artificially to mutate. Then the researchers sift through the resulting batch of randomly generated mutants looking for those mutants that best perform the desired task. Some amazing results have been achieved with this technique,[25] including:

- Improvement of enzyme activity.

- Increase of thermal and pH stability.

- Improvement of side activities.

- Improvement of stability against solvents and oxidants.

These are impressive achievements, but the technique does have its limitations:

- There must be a step-by-step mutational pathway to the new structure. The random walk doesn't do big leaps.

- One must be able to create a large enough mutant library in order to find the rare positive mutants.

- One must have a rapid screening method to detect the rare positive mutants.

More fundamentally, this laboratory approach is not like what we find in nature. It involves extremely high mutation rates, carefully chosen reaction conditions, the precise use of genetic engineering tools, and the artificial selection of variants. These are all hallmarks of design. The fact that this method cannot make big evolutionary leaps despite all of these intelligent inputs—inputs absent from any blind evolutionary process—is telling.

Designed changes. This name is only a misnomer to the degree that it implies that the alternative approach doesn't also involve design. The difference isn't design vs. no-design. Both involve design. It's that this approach, for one, dispenses with the use of randomness for generating a batch of mutants to sift through.

I have spent several years working with this second approach. I started to work with a xylanase enzyme in 1974, and since 1997 the aim of my research team was to modify its structure to improve its stability. We have designed disulfide bridges for the molecule in order to stabilize it against extreme temperatures and influence its pH stability and profile.[26] (To influence its profile means that there is a shift in the pH optimum. The enzyme that used to work optimally at, say, pH 7 can now work better at pH 8.) The bridges are formed automatically when two cysteine amino acids are in a correct position and distance from each other. So, of course, much of our planning work went into getting the two cysteine amino acids in place. Figure 10.6, row A shows part of the xylanase gene (upper row) and the respective amino acids in the enzyme (lower row). The positions modified by genetic methods are shown in bold. Figure 10.6, row B shows the two positions where the gene has been mutated. The mutations lead to incorporation of two cysteines (C; codon tgc), which replace threonine (T; codon acg/acc). A disulfide bridge is formed spontaneously between the two cysteine residues in the model structure shown in Figure 10.7.

The probability of forming one bridge randomly is very low—only one in 2×10^7 (20 million) The probability of forming two bridges through chance mutations is as low as one in 4×10^{14}. In practice this is completely out of reach for random methods. Many research groups have tried to improve xylanase stability by random methods and in some cases with good results, but they have never created disulfide bridges with this method. The probability of the formation of the disulfide

cag**acg**attcagcccggc...acgtac**acc**aatggtccc   **A**
Q **T** I Q P G... T Y **T** N G P

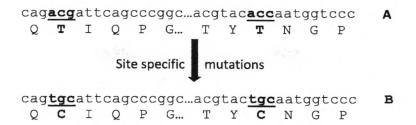

Site specific | mutations

cag**tgc**attcagcccggc...acgtac**tgc**aatggtccc   **B**
Q **C** I Q P G... T Y **C** N G P

FIGURE 10.6—A) Part of the xylanase gene (upper row) and the respective protein chain (lower row) before and B) after mutations.

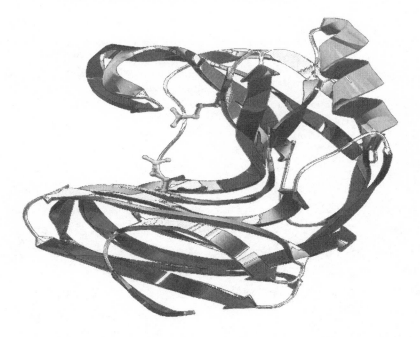

FIGURE 10.7—Molecular model of xylanase, which is a small enzyme that degrades xylan fiber (e.g. in birch wood).

bridge described above is actually much lower, because one has to change five nucleotides in the 669-nucleotide-long gene, which is one possibility in 1.4×10^{17}.[27] Even then you have only produced a somewhat more stable variant of the same structure.

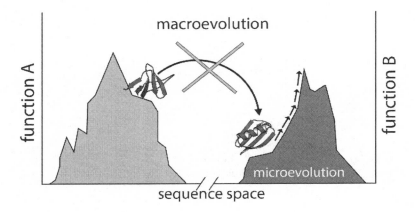

FIGURE 10.8—The difference between two protein structures in sequence space is represented by two mountains. Each mountain also represents the limits of microevolution. Each protein can vary modestly while maintaining its basic structure and activity, but if a random mutation moves a protein into the valley, it loses its function and structure and can travel no further on the path to the other protein "mountain."

Lehigh University biochemist Michael Behe's recent work provides some additional insight here. He surveyed the progress of mutations in large microbial populations over decades and then applied straightforward math to extrapolate to the maximum number of generations and mutations that might have occurred in the entire history of life on earth. He did this to offer an empirically rigorous estimate of what he refers to as "the edge of evolution," that is, the maximum number of simultaneous, coordinated random mutations that blind evolution is able to manage at any one point in the history of life. What, in other words, does the empirical evidence in this area say is the biggest leap evolution can make in a single mutational jump? Is it two simultaneous and coordinated mutations? Three? Ten? Fifty? Even more?

As we saw earlier, Behe looked at the decades-long experiment by Richard Lenski's team on *E. coli*. But Behe also studied, among other cases, the changes of the HIV virus and the malaria-causing protozoan *parasite* in the wild. Malaria is perhaps the exemplar of his study. There have been an estimated 6×10^{21} malaria parasite cells in the past sixty years, and during that time there have been some interesting malarial mutations. For instance, the malaria parasite developed resistance to the anti-malarial drug chloroquine. On the other hand, malaria has been utterly stymied, for thousands of years, by the sickle cell genetic mutation in humans. (See Chapter 4.)

If a person inherits the sickle cell mutation from both parents, he's horribly sick. But if the person gets a copy from only one parent, the person is fine and is resistant to malaria. Despite the staggering number of malaria parasites and the many generations of it over those thousands of years, the parasite has not been able to evolve a way around the defense conferred by the sickle cell mutation. Malaria also has not been able to evolve the ability to persist in cooler climates, even though it has had every opportunity to gradually evolve this capacity over the long ages of its existence.

So Behe studied the nature of the adaptive mutations that malaria successfully managed and, in particular, determined what these involved genetically. He also looked at what stymied malaria. He then compared these findings to insights culled from the study of HIV, *E. coli*, and other biological forms, and extrapolated to time-frames in the billions of years. From all this Behe was able to put a hard number on what the empirical evidence says is the "edge of evolution":

> The immediate, most important implication is that complexes with more than two different binding sites—ones that require three or more different kinds of proteins—are beyond the edge of evolution, past what is biologically reasonable to expect Darwinian evolution to have accomplished in all of life in all of the billions-year history of the world. The reasoning is straightforward. The odds of getting two independent things right are the multiple of the odds of getting each right by itself. So, other things being equal, the likelihood of developing two binding sites in a protein complex would be the square of the probability for getting one: ... 10^{20} x 10^{20}, which is 10^{40}. There have likely been fewer than 10^{40} cells in the world in the past four billion years, so the odds are against a single event of this variety in the history of life. It is biologically unreasonable.

... drawing the edge of evolution at complexes of three different kinds of cellular proteins means that the great majority of functional cellular features are across that line, not just the most intricate ones that command our attention such as the cilium and flagellum. Most proteins in the cell work as teams of a half dozen or more.[28]

My decades of experience with enzymes have left me with a similar view on the limits of evolution. Also, I followed the inevitable backlash after the publication of Behe's book on this subject, and found the usual brew of misreadings and strawman characterizations of his arguments. This is not how the defenders of a theory react when they have the evidential goods. That is the behavior of someone defending a theory in crisis.

Side-Deals and Sandboxes

MANY EVOLUTIONISTS recognize that the neo-Darwinian mechanism, traditionally understood, simply cannot make big mutational leaps, ones involving several simultaneous and tightly orchestrated genetic mutations. Having recognized this, they have gone looking for one or more patches to salvage evolutionary theory.

In the case of enzyme evolution, it has been suggested that new enzyme activities are formed so that the weak side activities are improved by random mutations.[29] As noted earlier, the side activity is just that— not the primary role of the given enzyme but a kind of minor bonus feature, if you will. A mutation that doesn't harm the enzyme's main activity but does modestly improve a side activity could represent something nature would select for. These slight improvements could thereby function as stepping stones on the way to larger and larger improvements, obviating the need for a sudden jump involving a cluster of simultaneous mutations.

This is, of course, logically possible, but there are a couple of problems with the idea. First, as noted earlier, these improved side activities come at the price of increased energy expenditure,[30] and experimental results show that nature does not select for slightly improved side activities at the cost of increased energy expenditure. Now, a skilled research

FIGURE 10.9—Molecular biologist Ann Gauger lecturing at Aalborg University in Copenhagen in November of 2012 on the topic "From the Problem of Metabolic Innovation to Principles of Design."

team overseeing an enzyme experiment could override nature and select for the weak side activity, but these are cases of intelligent selection, not natural selection.

Second, where slight improvements in side activities do occur, nothing new is formed; only the existing activity is strengthened while the basic protein structure remains the same. A recent paper by Axe and Gauger[31] emphasizes this point: Mutation and selection can improve good designs but never invent a design.

It also has been suggested that new proteins are formed via neutral or nearly neutral mutations. There is more than one option for where these neutral mutations could occur in the genome. For instance, some mutations involve a single DNA letter substitution that leaves a given function intact and doesn't obviously affect fitness positively or negatively. Also, a mutation may duplicate a stretch of genetic information, then subsequent mutations may occur in this duplicated stretch. In such a case the copied stretch is free to mutate to new functions without having to be functional each step along the way. Random mutations can conduct risk-free or neutral experiments in the duplicated sequence, and with luck, somewhere down the road all this no-risk mutational experimentation will lead to something reproductively useful, making it more likely to get passed on and spread.

This idea of mutations accumulating in an adaptively neutral zone before re-entering the survival-of-the-fittest arena has garnered excitement in some quarters of the evolutionary community, but as an explanation for the evolution of novel, complex, information-rich structures and organisms, the proposal is ultimately problematic: (1) It doesn't explain where the original gene came from, since the starting point involves a functional protein and doesn't build it from scratch. (2) A duplicated gene can change only within narrow limits, and proteins as a whole face this same limitation.

Work by Francisco Blanco at the European Molecular Biology Laboratory supports this second point.[32] Blanco and his team studied the sequence space between two different small proteins having different folds. One was a sixty-two-amino-acid protein that folds as an eight-stranded orthogonal β-sheet sandwich, and another was a fifty-seven-amino-acid protein that has a central α-helix packed against a four-stranded β-sheet. The authors designed a gradual series of mutants to discover whether there would be an evolutionary path from one fold to another. They concluded that the sequence space between the two proteins is enormous. The results suggested that only a small fraction of this space would have adequate properties for folding into a unique structure. The sequence spaces of the two small proteins did not overlap. Blanco concluded from this that getting a completely new protein from an existing one "is unlikely to occur by evolution through a route of folded intermediate sequences." In other words, he and his team do not see how a traditional neo-Darwinian pathway could make the journey.

Gauger and Axe performed a similar experiment, but chose two enzymes with structures that very closely resembled each other. They made twenty-nine specific amino acid changes to one of the enzymes without being able to change its function to another one. They concluded that even this modest attempted change would require at least seven nucleotide changes and, considering the known mutation rates, at least 10^{27} years, which is longer than the estimated age of the earth. They concluded that "this result and others like it challenge the conventional practice of inferring from similarity alone that transitions to new functions occurred by Darwinian evolution."[33]

I applied in 2006 for financing for a project from the Finnish Academy for a similar project. I wanted to find out whether two closely related enzymes (family 11 xylanase and endoglucanase III) of *Trichoderma* fungus can be mutated from one to the other through a series of small, genetic mutations. I did not get the funding. According to the expert panel, I would probably not succeed and no one knows how enzyme structures are formed. I was disappointed not to get the funding, but I did find refreshing the panel's frank admission that the neo-Darwinian mechanism does not provide a credible scenario for the evolution of enzymes.

Douglas Axe's subsequent work strongly suggests that what Blanco's team demonstrated also spells trouble for neutral evolution. Neutral evolution gets around the problem of needing functional protein forms at each mutational step. The mutations are in a genetic neutral zone, after all; mutations don't need to improve fitness every step of the way there. But this advantage comes at a steep price: The neutral zone trades away the power of natural selection to guide the journey from one protein form to another. Remember, it was the joint mechanism of random variation *and* natural selection that made Darwin's theory appear so plausible. Natural selection, working on random variations, was said to function as a designer substitute. Without natural selection, the neutral zone experiment in genetic mutations is blind even to mutations that would confer an immediate advantage outside the neutral zone. That is a huge price to pay, because biologists have long known that a purely random walk through sequence space could not overcome the long odds to evolve all the biological complexity around us. That's precisely why they champion, and have clung so ardently, to neo-Darwinism: natural selection was believed to be the guide and savior of random variation, giving it traction and direction. Without it, the random walk is blind, drunk, and without even an immediate purpose.

So there appear to be strict limits to how far proteins, including enzymes, can evolve. We needn't view this as a failure of these biological forms. The truth is just the opposite. Bloom et al.[34] have shown that extra stability allows more mutations and makes the protein more flexible within narrow limits without destroying its natural structure. In other words, the protein scaffold can tolerate an impressive number of mutations without change in its basic structure.

I have now briefly reviewed the key results obtained during the last three decades of protein—and especially enzyme—engineering. The results can be summarized as follows:

+ Proteins can be modified with random and design methods—but only within narrow limits: basic structures have not been changed.

+ Although randomness plays a role in some of the protein experiments, all the experiments are designed and, thanks to the intelligent input of the experimenters, have searched much larger space than natural processes could have searched.

+ Even with huge amount of intelligent input, nothing fundamentally novel has been created.

And what about the beginning of life, and with it, the beginning of proteins and enzymes? Remember that even if all of the above problems could—against all the evidence—be overcome by some blind evolutionary process, materialistic evolution would still face an insurmountable challenge, one that can be summarized briefly: *Enzymes are biochemical machines crucial to life. These proteins catalyze all the reactions in the cell. They recognize, cut, glue, transport, oxidize, move, and change parts of molecules. But how do you get enzymes, or any kind of protein, in the first place?* Biologist Dan Tawfik of the Weizman Institute in Israel oversees a research group dedicated to finding pathways by which proteins may have evolved. But he is candid about the origin-of-life problem. "Evolution has this catch-22: Nothing evolves unless it already exists," he said. So what does he make of the origin of the first enzymes and other proteins, essential ingredients for life? He described their origin as "something like close to miracle."[35]

11. THE CHASM WIDENS

In 1847 a Hungarian medical doctor, Ignaz Semmelweis, noticed that many women died after childbirth of childbed fever. After investigating the situation more closely he started to suspect that this was somehow connected to doctors who came directly from autopsy to examine the women after birth. Semmelweis suspected that the doctors brought something on their hands that caused the childbed fever. His suspicions intensified when one of his colleagues got similar symptoms after cutting his finger during an autopsy. After he ordered doctors to wash their hands with chlorine water, the patient death rate plummeted.

In spite of this clear evidence that Semmelweis was on to something, his colleagues and the larger scientific community did not take him seriously. He was humiliated and, growing increasingly bitter and erratic, was lured by a colleague into an insane asylum. When he tried to leave, he was beaten so severely by the guards that he died a few days afterward.

Years later Semmelweis was vindicated when the role of bacteria in disease became clear. The term "the Semmelweis reflex" was coined in honor of him and as a warning to scientists. It refers to "the rejection of new knowledge because it contradicts entrenched norms, beliefs and paradigms."[1]

The history of science shows that the Semmelweis reflex is all too common among scientists. When the evidence turns against a scientific paradigm, it often dies only very slowly and painfully, be it the geocentric cosmology, the more recent static universe cosmology, humoralism, unmoving continents, phlogiston, or neo-Darwinism.

The parallels between the now-defunct theory of phlogiston and neo-Darwinism are particularly illuminating. Of phlogiston, Antoine Lavoisier, the father of modern chemistry, offered this observation:

> Chemists have made phlogiston a vague principle, which is not strictly defined and which consequently fits all the explanations demanded of it. Sometimes it has weight, sometimes it has not; sometimes it is free fire, sometimes it is fire combined with an earth; sometimes it passes through the pores of vessels, sometimes they are impenetrable to it. It explains at once... transparency and opacity, color and the absence of colors. It is a veritable Proteus that changes its form every instant![2]

Phlogiston was foundational to chemistry education from the fifteenth to the end of the sixteenth century, although the problems associated with it were observed a century and a half before it even started to waver. The theory held that a mysterious substance called phlogiston was released from a burning substance, but already in 1630 James Ray wondered why the oxide of tin was heavier than the starting material if burning had released phlogiston. But not to worry: The supporters of phlogiston theory reasoned that in some cases phlogiston could have a negative weight![3]

The story of phlogiston shows how an established paradigm may persist in the face of contrary evidence because its supporters patch it up *ad nauseam* instead of following the evidence.

The Darwinian theory of evolution is the phlogiston of our day, festooned with a myriad and growing number of patches. Evolution is slow and gradual, except when it's fast. It is dynamic and creates huge changes over time, except when it keeps everything the same for millions of years. It explains both extreme complexity and elegant simplicity. It tells us how birds learned to fly and how some lost that ability. Evolution made cheetahs fast and turtles slow. Some creatures it made big and others small; some gloriously beautiful, and some boringly grey. It forced fish to walk and walking animals to return to the sea. It diverges except when it converges; it produces exquisitely fine-tuned designs except when it produces junk. Evolution is random and without direction except when it moves towards a target. Life under evolution is a cruel battlefield except when it demonstrates altruism. Evolution explains virtues and vice, love and hate, religion and atheism. And it does all this with a growing number of ancillary hypotheses. Modern evolutionary theory is the Rube Goldberg of theoretical constructs. And what is the result of all this speculative ingenuity? Like the defunct theory of phlogiston, it explains everything without explaining anything well.

Fred Hoyle and Chandra Wickramasinghe urge us to be cautious of a theory that needs ever more ancillary hypotheses when faced with new facts:

> Our interpretation is:
>
> 'Be suspicious of a theory if more and more hypotheses are needed to support it as new facts become available, or as new considerations are brought to bear.'
>
> This interpretation leaves Darwinism in a poor way, for this is exactly what has happened to Darwin's theory.[4]

The central problems of the theory of evolution have been known from the beginning. With growing scientific knowledge, the problems have only increased. Darwinism turned to neo-Darwinism and now we are in a post-neo-Darwinian era, even as biologists wedded to a designer-free theory offer up various new and ancillary hypotheses to save the idea of blind evolution. In search of a solution, Richard Goldschmidt subscribed to an old idea called saltation[5] (evolution by jumps). Saltation would explain why there are no links (living or fossilized) between basic animal groups, but Goldschmidt was unable to offer an explanation for how new forms could emerge so quickly. His view was knocked down as biologically impossible. In 1972, Stephen Jay Gould[6] and Niles Eldredge noted the fossil record's persistent pattern of abrupt appearance of novel biological forms followed by long periods of stasis. They proclaimed neo-Darwinism dead and suggested punctuated equilibrium to replace it. But biologists rejected the idea for lacking a credible mechanism for generating biological change so quickly, and later Gould adjusted his theory toward neo-Darwinism.

But efforts to find an alternative to neo-Darwinism persisted. Of particular note, in summer 2008 sixteen researchers representing evolutionary biology, paleontology, and philosophy met at the Konrad Lorenz Institute in Altenberg, Austria. Media called the group the Altenberg 16.[7] The members of the group had different and even contradictory views, but all agreed that the modern evolutionary synthesis was in trouble, and that new hypotheses were needed to explain the origin of biological form. Graham Budd, one of the participants, said, "When the public thinks about evolution, they think about the origin of wings and the invasion of the land... But these are things that evolutionary theory has told us little about."[8]

The Altenberg 16's effort to break neo-Darwinism's stranglehold on origins biology is a breath of fresh air. The development of science thrives on open competition among different models. Without a willingness to consider an alternative model, a wrong assumption in a dominant model may persist in the face of mounting experimental evidence, since the proponents of the model will either attribute the experimental results to error or add an ancillary patch to explain away the results.

A recent example is the static-eternal model of the universe. Albert Einstein was so wedded to it that he built a fudge factor into his equations after he realized that his general theory of relativity appeared to undermine it. The fudge factor helped explain away the unwelcome implication. But while reluctant, Einstein and some other influential physicists and astronomers were willing to consider a competing model soon propounded by astronomer Edwin Hubble, according to which the universe had a beginning and was expanding. This, in combination with mounting experimental evidence in favor of this new model, eventually doomed the static-eternal universe model.

As the above example illustrates, a new model that better explains the data is typically needed before scientists will, *en masse*, abandon the once dominant paradigm. If all scientists were strictly rational and unbiased, a competing theory wouldn't be necessary to doom an old paradigm, but scientists are human after all—a fact well attested by both biology and the history of science.

A Worldview Out Every Window

SCIENTISTS MAY seem a proud lot, but even the proudest of scientists needs a good dose of humility if he is to be guided by the evidence rather than by personal biases. One way to cultivate this attitude of humble flexibility before the evidence is to recognize that every scientific model that touches on origins automatically also has worldview implications. In this respect naturalism is comparable to theism. Supporters of intelligent design are often accused of mixing a worldview (theism) with science. But a similar charge could be applied to those touting origins theories consonant with atheism. For example, the unprovable idea known as the multiverse has been offered to explain away the curious fact that the laws and constants of physics and chemistry appear fine-tuned to allow for stars, planets and, with them, life somewhere in the universe. On the multiverse theory, there are innumerable other universes and ours is just one of the lucky ones with the right parameters for life—one of the minority of universes that won the proverbial lottery, if you will. The multiverse model, its adherents insist, removes the need to invoke a designer—a fine-tuner—to explain fine-tuning. But notice that this means the theory has clear worldview implications. So the multiverse theory should not be allowed, right? Or is it only scientific theories with theistic implications that are not allowed?

The same issue arises with the origin-of-life question. Was the first microscopic life on earth the result of blind forces and luck, or was intelligent design involved? Either hypothesis has worldview implications, but for the scientist, the relevant question should be: Which explanation is best supported by reason and physical evidence?

The theory of intelligent design holds that the appearance of design in nature is real, rather than illusory, that living organisms are sophisticated information systems that are best explained as being the result of an intelligent cause. The founders of modern science were convinced that nature pointed to design, and a growing number of contemporary scientists think this way as well.

Even some religiously skeptical philosophers have begun to entertain the possibility of intelligent design. The well-known philosopher Antony Flew (1923–2010) defended atheism for almost all his life. However, in 2004 he changed his mind and gave this for the reason: "The argument to Intelligent Design is enormously stronger than it was when I first met it."[9] New York University philosopher Thomas Nagel—an atheist—endorsed Stephen Meyer's book *Signature in the Cell: DNA and the Evidence for Intelligent Design*,[10] and himself published a book in 2012 with the pointed title *Mind & Cosmos: Why the Materialist Neo-Darwinian Conception of Nature Is Almost Certainly False*. There he writes that design theorists such as Michael Behe and Stephen Meyer have offered "empirical arguments… of great interest," and should not be simply reviled and dismissed. "Even if one is not drawn to the alternative of an explanation by the actions of a designer, the problems that these iconoclasts pose for

FIGURE 11.1—As one newspaper article about me noted, "Criticism of evolutionary theory is a stressful hobby." It's brought me a lot of criticism, as I pointed out at the time: "Every time I've been suggested for a new position, a question mark has been placed behind my name. On the other hand, life as a dissenter is rich and exciting." (Simopekka Virkkula, "Suomalaisprofessori haastaa Darwinin luoman kehitysopin," *Aamulehti*, July 20, 1997.)

the orthodox scientific consensus should be taken seriously. They do not deserve the scorn with which they are commonly met. It is manifestly unfair."[11]

Yet some evolutionary materialists, in the face of all this, have persisted in caricaturing intelligent design. University of Jyväskylä philosopher professor Tapio Puolimatka wrote an article for Finland's leading newspaper, HS, to highlight Nagel's views on the matter.[12] The article, entitled "The Theory of Evolution Must Be Taught in a Critically Open Way," was immediately attacked as creationist/fundamentalist propaganda. Puolimatka described the situation in his response article:

> My article mainly consisted of the thoughts of Jewish atheist philosopher Thomas Nagel. I presented the analysis of an atheist philosopher, of how basic beliefs influence the interpretation of facts. It seems to have been difficult for people to accept Nagel's thought that the only way to teach biological facts in a neutral way is to admit that evidence can be interpreted in different ways and this can lead to different conclusions depending on which religious starting point has been used in the interpretation. According to Nagel, his own atheistic conviction that there is no God, and the theistic belief in an almighty and all-knowing God, are both the same type of basic beliefs... Professor Valtaoja... criticizes this view. According to him this claim means giving up the successful basic method of science only to replace it with a "fundamentalist type God." It was bewildering to see that the thoughts of a Jewish atheist [Nagel] were considered as Christian creationism and fundamentalism.

It is also notable that Esko Valtaoja apparently does not see his own naturalistic basic beliefs. Valtaoja accuses Puolimatka of mixing religion and science, but Valtaoja is mixing his naturalistic beliefs with science without even noticing it. And whereas we seek to base our scientific case for design on reason and physical evidence, Valtaoja appears eager to win the argument by a dogmatic appeal to naturalism.

Another approach to countering intelligent design is to argue that, in essence, God wouldn't have done it that way. This at least has the virtue of offering an argument rather than merely a question-begging methodological rule. Professor of evolutionary ecology Hanna Kokko, and Katja Bargum, the science editor of Finland's national public service broadcasting company, Yle, argue for evolution by insisting that an intelligent designer would not design certain biological structures which they think are stupid.[13] This is a common line of attack among ID opponents, but many of their "bad design" examples collapse under scrutiny. For example, the supposed "backward wiring" of the vertebrate eye actually improves oxygen flow, and organs deemed vestigial and useless have proven to perform valuable functions. But more basically, these bad-design arguments rest on a suspect theological assumption, namely that if there is a God, he would design every organism to be maximally fit and free of pain or weakness—every creature a little god.

The problem with this reasoning is that there are well-established theological reasons why a good and wise God would not create such a world, particularly one he knew would be peopled by fallen and sinful humans. Anti-design evolutionists ignore this rich body of theological reflection, invoke a superficial theology of creation, and then trash the strawman as incompatible with evidence from biology. And if you call them on this, you are accused of talking theology in a science discussion. They deserve credit for brazenness, at least, since they are the ones who introduced theology into the discussion, and badly at that.[14]

Meanwhile, championing atheism from the university lectern gets a pass. Professor Stephen Jay Gould lectured in Helsinki University in 1999, and no one opposed his visit although he had been very clear about his religious views: "No intervening spirit watches lovingly over the affairs of nature," he had commented. "No vital forces propel evolutionary change."[15] Richard Dawkins lectured in Finland in 2005, and no petition was gathered against him, although he freely mixes religion and science. He even wrote a book called *The God Delusion*.[16] Physics professor Kari Enqvist teaches at Helsinki University and openly mixes religion and science while saying that "faith in God is like a viral disease."[17] While I believe Enqvist is wrong, I support his right to speak his mind. This is called academic freedom. Unfortunately, it's a scarce commodity these days. We don't need less of it. We need more of it—both for those who think science points to atheism, and for those of us who disagree.

Faith Inescapable

SOME ATHEISTS frame the debate as faith vs. reason, but that's a muddled—indeed, unreasonable—way to frame the controversy. In autumn 1987 I was sitting in the office of biochemistry professor Kaspar Winterhalter in Zürich. I had applied for a teaching position, and we were writing a joint publication on enzyme characterization. We had the following discussion:

"Doctor Leisola, you are a very religious man!"

"Professor Winterhalter, so are you!"

"What do you mean?"

"Your world view, like mine, is based on things that cannot be proved but have to be accepted finally by faith."

"Hmm... you may be right."

We both had a presupposition about the nature of reality. Each of us thought that his own view was reasonable and fit the facts. But neither of us as finite men could prove with mathematical certainty his own starting point. In this respect we both were believers.

Understand, my point isn't that our two views are wholly equal or necessarily irrational. I am convinced that the evidence for the intelligent design of nature is far stronger and more reasonable than the alternative. It's simply that both views, at the end of the day, reach beyond the seen to the unseen, and each of us trusts in something that cannot be proved in the way one might prove that Person Y is hiding in the closet, or that a square one meter wide has a circumference of four meters.

Given all this, it is misguided and unfortunate that in many universities one can freely use theological arguments mixed with science to speak for atheism, while scientific arguments that count in favor of theism are considered to be insulting and bad for the reputation of the university.

Darwinists in Denial

WHILE WORKING on an earlier draft of this book, I was reading the manuscript of Stephen Meyer's *Darwin's Doubt: The Explosive Origin of Animal Life and the Case for Intelligent Design*, whose first fourteen chapters explore the problems with neo-Darwinism. After noting that the evidence continues to accumulate against the theory, Meyer comments, "Nevertheless, popular defences of the theory continue apace, rarely if ever acknowledging the growing body of critical scientific opinion about the standing of the theory. Rarely has there been such a great disparity between the popular perception of a theory and its actual standing in the relevant peer-reviewed scientific literature."[18] In the journal *Nature* Philip Ball admits much the same thing:

> We do not know what most of our DNA does, nor how, or to what extent it governs traits. In other words, we do not fully understand how evolution works at the molecular level... Yet, while specialists debate what the latest findings mean, the rhetoric of popular discussions of DNA, genomics and evolution remains largely unchanged, and the public continues to be fed assurances that DNA is as solipsistic a blueprint as always.[19]

This admission is particularly notable given the journal's steadfast commitment to naturalism, and given the central role neo-Darwinism plays in propping up naturalism.

As earlier noted, evolutionists long insisted that only a small part of our genome is useful, the rest being junk from evolution's long and wasteful trial-and-error process. This was conventional wisdom in the biological community, but already in the 1980s I viewed this talk about junk DNA as itself junk, and I discussed the topic with scientists in Switzerland. More recently, even just fifteen years ago, the pro-Darwinian molecular biologists I knew thought that most of the secrets of the genome had already been discovered. The situation in one short decade has reversed, and the shift has moved in the direction of design theory.

Research motivated by the design paradigm has always assumed that our genome contains relatively little that has no purpose. This also led pro-design biologists to assume that there was much still to be discovered about the genome. Both design-theoretic expectations have proven true while the contrary neo-Darwinian expectations have proven false.

We are in the midst of a genomic gold rush, an exciting race to see who can uncover the next intriguing function of this or that vein of genetic information formerly deemed junk.[20] At the same time, protein-coding genes—genes long understood not to be junk—are doing much more than previously thought. For example, a *Nature* article reveals that a yeast genome with 6,000 genes can produce hundreds of thousands of different messages, depending on how the genes are read.[21] In the light of these discoveries, it is almost humorous to read claims that design theory is not a useful paradigm for research, and that without evolution biology does not make sense.

What's next? We are in the middle of an exciting paradigm shift, but old and dominant paradigms die slowly, and there is more reason than usual for this to hold in the case of evolution. Recall the remark by atheist Michael Ruse quoted earlier: "Evolution is promoted by its practitioners as more than mere science. Evolution is promulgated as an ideology, a secular religion—a full-fledged alternative to Christianity... Evolution is a religion. This was true of evolution in the beginning, and it is true of evolution still today."[22]

New Evolutionary Hypotheses

A GROWING number of biologists have in practice given up the traditional neo-Darwinian theory. Already by 1980 the late Harvard paleontologist Stephen Jay Gould proclaimed that neo-Darwinism was dead. But in an academic world controlled by naturalism, this has not led to the death of evolutionary theory but to a scramble to prop up it up with various ancillary fixes. We turn now to a brief overview of some of these efforts.

We have already looked at punctuated equilibrium and neutral/non-adaptive evolution. Here are some others:

Self-organization. In the 1990s a group of scientists at the Santa Fe Institute in New Mexico developed a hypothesis they called self-organization,[23] meant to explain the origin of biological systems strictly by reference to chemical and physical laws and processes. Its proponents point to the spontaneous emergence of order from disorder in nature, such as crystals or the spirals of hurricanes or chambered nautiluses. This approach works well to explain mathematically compressible forms of order—ones that can be expressed in an algorithm. But it does not explain the aperiodic, non-compressible order that is biological information.

The difference is easily illustrated. The string of letters acegikmoq-suwyac... follows a strict pattern that could be described by a brief algorithm, one that if plugged into a computer could churn out a long string of letters. But notice how different that pattern of letters is from the pattern of letters on this page, or the letters in an instruction manual, or the letters and other symbols in a software program. The latter examples are all aperiodic and do not follow a rigid algorithm throughout. Genetic information is this kind of order, and proponents of self-organization have not been able to offer examples of novel biological information self-organizing in the present, nor a plausible scenario by which it might have happened in the past. Algorithms may generate beautiful patterns. They do not generate novels or software programs or the reams of biological information needed to code for novel biological forms.[24]

Evo-Devo. Proponents of evolutionary developmental biology (evo-devo) hold that understanding how organisms develop from embryos will shed light on how they evolved. They draw inspiration from the insight that genetic information expressed early in embryonic development tends to have a major influence on the basic structure of an organism. This, in other words, seems to be where the action is in terms of big, splashy morphological innovations. In particular, some developmental biologists have emphasized homeotic genes (including *Hox genes*), which regulate development of key anatomical structures in various organisms.[25] Evo-devo proponents have suggested that mutations to such regulatory genes may have caused changes in basic structures, allowing the evolutionary process to generate novel forms much more rapidly than previously assumed.

But after some initial excitement, the evo-devo hypothesis has run into trouble. Research stretching over several decades confirms that mutations in regulatory genes are generally disastrous for the mutated organism. In 2014 evo-devo proponent Wallace Arthur admitted this. He expressed a continued faith in common descent and in the evo-devo project, but also confessed that "when the fitness consequences of large-effect mutations in early development have been studied, they have in almost all cases been found to result in major fitness decreases. This is true, for example, of the homeotic mutations studied in *Drosophila* [fruit flies], and it is one of the main reasons for the rejection of Goldschmidt's saltational theory of evolution."[26]

Epigenetic inheritance. In recent years we have seen that part of a cell's information is outside DNA. Could this type of information—called epigenetic information—have evolutionary consequences? Epigenetic information can be influenced from outside and be inherited by the next generation without changes in DNA.

Eva Jablonka from the University of Tel Aviv defends a new view of evolution that contains elements of such evolutionary change, a view that does not mesh with neo-Darwinism.[27] Jablonka has collected evidence for her view of an epigenetic inheritance system, and says that changes in metabolism caused by the environment can be inherited without changes in DNA. She emphasizes that much structural information that is responsible for the form of an organism is inherited from the parents independent of DNA—for instance, via membranes and other three-dimensional structures in the cell. Chemical changes to DNA that do not change its nucleotide sequence (like methylation) may have an effect on the regulation of genes. She mentions a recently discovered RNA-mediated epigenetic inheritance.[28] Small RNAs together with enzymes affect gene expression and the structure of chromatins. The mechanisms cited by Jablonka, however, do not explain macroevolution. She is forced to conclude that "since, with few exceptions, the incorporation of epigenetic inheritance and epigenetic control mechanisms into evolutionary models and empirical studies is still rare, our discussion is, inevitably, somewhat speculative."

Natural genetic engineering. An earlier chapter looked at the work of University of Chicago geneticist James A. Shapiro who, together with Richard von Sternberg, published articles critical of the modern evolutionary synthesis. Shapiro calls his view natural genetic engineering: Organisms modify their genomes as a result of changes in the environment.[29] He shows that such changes are not random. Mutations seem to be regulated and the organisms seem to react intelligently to the environment. He refers to a bacterial SOS-system activated as a result of DNA damage. The cell starts to produce DNA polymerizing enzymes that make mistakes. The system is harmful for the organism but produces mutations that lead to damage repair. Once the damage is under control, the error-prone polymerase is inhibited.

Shapiro's work is fascinating and opens up new aspects of the cell's information systems. However, just how this type of programmed ability to react to the environment emerged remains unanswered. It is a sophisticated system whose origin itself cries out for an explanation, but none is forthcoming. The observations of Shapiro make the cell actually more complex than previously thought and thus a bigger challenge to any unguided evolutionary process.

These are some of the more prominent patch theories, put forward in response to experimental data but also in hopes of rescuing modern evolutionary theory. There are many others, including hybrids.[30] There is talk of biological control systems in evolution, of postmodern evolution. There are whispered hopes of discovering some way to reintroduce saltation (abrupt evolutionary leaps). As naturalists, the researchers pushing these hypotheses do not doubt evolution as such. For many of them, that is not allowed. They only doubt the neo-Darwinian mechanism. But all of the proposed patches come with one or more fatal limitations—and in every case the limitations can be boiled down to an inability to generate novel and advantageous biological form and information. For that we must look to a very different type of cause, one now in operation and with the demonstrated power to generate novel form and information—intelligent design.

12. THROUGH A DOORWAY TO ADVENTURE

THE CONFRONTATION IN ORIGINS SCIENCE, FINALLY, IS NOT BEtween science and religion, but between a mindless process and a process intelligently guided. Either the world's order and its living forms arrived blindly, and blindly unfolded through the laws of chemistry and physics, or they came from a designer who made these laws and forms. Man has, however, developed a third option—a synthesis of the blind and guided approaches. It often goes by the name theistic evolution, though theistic Darwinism would be a clearer label, since we have in mind here those who attribute nature to an omnipotent maker while seeking to preserve the blind watchmaker of neo-Darwinism. On this view, God spoke the cosmos and its laws and constants into existence at the Big Bang, but he created the diversity of life through secondary causation—namely by using purely random mutations yoked to natural selection, which chose the strong and destroyed the weak.

Some theologians find themselves attracted to this hybrid approach, and I understand how it might be a tempting option for those repeatedly told that evolution is a "fact" supported by "the scientific consensus." I understand because I myself was convinced of it in this way as a young scientist. But my journey from Darwin to design has convinced me that the great weight of scientific evidence is against theistic evolution[1] because it is against blind evolution generally. The biological evidence, taken carefully and in total, does not point to the evolution of all living things from common ancestors by unguided processes. It points away from this.

When I started my studies in 1966, it seemed that the doctrine of evolution (which then was essentially neo-Darwinism) was an eternal scientific truth. But already in the 1970s there was stubbornly contrarian evidence I was forced to grapple with, and the last thirty years of research has revealed many other things that do not fit neo-Darwinism. This is why suspicion of the theory has spread in spite of the massive propaganda of leading science organizations, journals, and the popular media. The telling part of this is that it has spread not only among what are termed design theorists, but also among biologists who remain eager to keep a designer out of origins science, even as they remain without an adequate creative mechanism to explain the origin and diversity of life.

I have seen enough, however, to know that evolutionary theory will not go gently into the night. People do not easily give up the views adopted as a part of their education and culture. There is also a lot at stake: reputation, money, power, and for some, a worldview with its lifestyle. For many it's as simple as not wanting to be isolated and ostracized. This is an understandable fear, even if succumbing to it is surely short of the heroic. I would merely urge those who at some level know better to consider taking the hard path, and not merely because professional courage is a virtue worth pursuing. There is more at stake.

Once a high school teacher gave me an essay by one of her students. According to the student, science had given him reasons to believe that life is meaningless. The student's situation saddened me. All too many students fail to differentiate between scientific results and their philosophical interpretation. The philosophy of materialism does point to an ultimately meaningless universe, but science doesn't. Scientific discoveries in a variety of fields point to a cosmos, not a chaos. They point to a universe bursting with evidence of meaning and purpose. Unfortunately, science journals and the popular media have long been preaching materialism's gloomy message under the guise of science, so it is not surprising that many students swallow it whole.

Years later, with that student's despairing letter still knocking around in my head, I wrote the following column for my university's newsletter:

Worldview Footprint?

Scientists do not function without worldview commitments, and their worldview easily affects the interpretation of their research results. These interpretations can and often will influence the worldviews of the members of the society. Viktor Frankl was a professor in the medical faculty of Vienna. As a Jew he was sent to one of Nazi Germany's concentration camps, Auschwitz, but survived. Frankl was "absolutely convinced that the gas chambers of Auschwitz, Treblinka, and Maidanek were ultimately prepared not in some Ministry or other in Berlin, but rather at the desks and in the lecture halls of nihilistic scientists and philosophers." The Nazi regime did not force scientists to work for them but "many scientists voluntarily oriented their work to fit the regime's policies—as a way of getting money... Most researchers, it turns out, seem to have regarded the regime not as a threat, but as an opportunity for their research ambitions" ("Uncomfortable Truths," *Nature* 434, no. 7034).

Professor Ernst Haeckel had already, before the First World War, laid the foundation for the Nazis' racist views, which were generally accepted by the science community. The father of Finnish genetics, Harry Federley, corresponded with Haeckel. He embraced racism and lectured in the world's first Eugenics Institute in Sweden. In Finland Federley pushed through the sterilization laws for criminals and the mentally handicapped. The laws were in force till 1970. Haeckel and Federley were monists (matter is the only reality) and had an enormous influence on society (*Jahresbuch Europäisches Wissenschaftskultur* 2005, 1:1).

The shadow of their worldview hangs still above our culture. It was recently expressed in the school shootings (Kauhava and Jokela) where the motive was the principle of natural selection to eliminate the despised. The shooters were victims of the teachings of our culture. Young people tend to be more radical (the word comes from Latin and means going to the root) and function on the basis of their beliefs. Luckily, not all naturalists are that consistent.

We rarely think that as university teachers we have to bear the responsibility of the worldview we communicate to students. But the university law obliges us to educate the youth to serve homeland and humanity. Therefore we teachers should recognize the faith commitments of our own worldviews and be careful how we communicate them to students. Fifteen years ago a teacher gave me an essay of a 15-year-old boy: "I studied science journals and formulated a solid worldview for myself. There is no God, no spirit, no meaning. It does not matter if I die now or after fifty years." It is frightening to think that my own teaching might leave this kind of a footprint and even more frightening to think where it might lead.

I got much feedback on that column, including the following: "First time during our long career in the university we could read an article from the weekly newsletter that forced us to think! For us humanists the topic is close to our heart and we hope that technically oriented members of our community would think about your words."

The evolutionists cannot both have their cake and eat it, at least not in any realm where reason and common sense hold sway. That is, it's unreasonable to argue for a blind watchmaker *and* to insist that arguing for a seeing watchmaker is *verboten*. Evolutionary theory speaks to the question of biological and human origins, and the question is worth grappling with fairly, both for its scientific import and for the issue's wider philosophical implications.

It's one that has occupied the minds of great thinkers in every age of the West. Recall a quotation from Chapter 1, a passage from the *Philebus* of Plato (427–347 B.C.E.). In it Socrates asks the key question: "whether we are to affirm that all existing things, and this fair scene which we call the Universe, are governed by the influence of the irrational, the random, and the mere chance; or, on the contrary, as our predecessors affirmed, are kept in their course by the control of mind and a certain wonderful regulating intelligence." Ever since then, great thinkers have debated those two possibilities. It's educationally backward to declare this monumental issue off limits and insist that a properly rigorous approach to origins may only entertain the materialist position.

In our day, design theorists employ reason and recent scientific discoveries to strengthen the case for design. As part of that argument, we often point out that investigators use design detection in archaeology, SETI research, cryptography, and various other scientific fields, and that humans rightly detect design every day without even thinking about it. When I speak of design detection, I often use a toothpick as an example. Everyone in the audience agrees: Natural process cannot produce a nicely formed toothpick. Clearly the toothpick was designed. Why then do many scientists either avoid the topic or get angry with scientists such as Michael Behe for suggesting that irreducibly complex systems in biology are best explained by reference to intelligent design? Many even do so in cases where "the miracle of natural selection" cannot be invoked, such as the fine-tuning of the universe or the origin of the first self-reproducing organism. The reason for such reactions is that intelligent design challenges a basic belief system of some. A challenge on this level can seem to shake the foundations of one's being, so it's no wonder that some people get nervous.

While discussing natural selection, evolutionary biologist Graham Bell commented, "A light bulb or a lathe are prefigured in the mind, and constructed according to a plan. It is entirely reasonable to assume that beetles and daises must be constructed after the same fashion, especially since they are much more complicated than anything that human ingenuity has so far managed to devise."[2] He agrees that a conclusion of design is perfectly reasonable and is based on observation; he simply thinks that Darwin's answer is a better one. I disagree with that last part: I find Darwin's answer a poor one. Now, science cannot decide the answer in the same definitive way one might solve, say, a tricky mathematical equation, but science can point us in the right direction. Reason and evidence can guide us, provisionally at least, toward the best explanation, but only if we agree to follow reason and attend to the evidence, rather than dogmatically ruling out one possible explanation.

Of Code and Coders

WE HAVE covered a lot of ground in the chapters of this book. In doing so it's easy to miss the forest for the trees, to miss what is key. One key moment for the origins debate was the discovery and elucidation of biological information, and at precisely the time when the science of information was coming into its own. Information has become one of the central research topics of our day. We speak of an information society even as molecular biology has become a science of information.

In 2012 the results of ENCODE were published and showed that the great majority of DNA is not junk but functional. The ENCODE results confirmed and extended what has grown increasingly clear ever since Watson and Crick first unraveled the double-helical structure of DNA more than sixty years ago: Far from being a cobbled-together, trial-and-error hack job, the cell is the most sophisticated information system known to man. The expectations of ID scientists were right.

Where does one go from here? Any time we encounter coded information and can trace it back to its origin, it always leads to a coder, to a designing intelligence. Increasingly I am convinced that only philosophical presuppositions prevent a person from detecting the intelligence behind the huge information content of life.

Expanding Mysteries

To BE sure, when we detect design, other questions quickly emerge. Who carved the text on the famous Rosetta Stone, which allowed historians finally to unravel Egyptian hieroglyphics? Who were the particular engineers who designed the equally famous rock formation known as Stonehenge, and how did they haul those massive stone slabs over such long distances? We do not know, although we can make good guesses. How specifically did the author of DNA produce it? Again, we do not know. And yet, as with Stonehenge and the Rosetta Stone, design detection remains possible.

Even in the case of contemporary productions of information, we understand less than one might suppose. How did Leisola and Witt write

FIGURE 12.1—The Rosetta Stone.

this book? We each used laptop computers and shared drafts via the internet. But that explanation only takes us so far. Somehow an author's thoughts are transmitted from the brain via nerve cells and through the fingers to the computer. But inescapable here is the mystery of consciousness, of mind, of choice. One may seek to reduce the mystery to some mechanism or mechanisms, but to reduce choice to mechanism is to obliterate choice. There may be a mechanism for actualizing a choice, but if a mechanism is responsible for the choice, then there was no choice. This won't do.

We need not embrace a simplistic dualism that sees no influencing role for the body and brain on the mind to recognize that mind, along with our awareness of freedom, remains the most primary thing we experience. Nothing is more immediately, more intimately evident, so it doesn't do to dismiss it as an illusion. Who, after all, is experiencing the illusion? And if choice is an illusion foisted on us by a blind process of evolution, then why should we trust our reason at all—the very reasoning faculty that supposedly tells us that we evolved through a blind process? Evolutionary materialism saws off the very branch of reason that it sits on.

The I—the mind—is inescapable. Naturalistic science has no answer to this mystery of mysteries. "The existence of consciousness is both one of the most familiar and one of the most astounding things about the world," writes atheist and philosopher Thomas Nagel.[3] Cognitive scientist and philosopher Jerry Fodor goes further. "Nobody has the slightest idea how anything material could be conscious," he remarks. "Nobody even knows what it would be like to have the slightest idea about how anything material could be conscious."[4]

Atheist Richard Dawkins also concedes this. "There are aspects of human subjective consciousness that are deeply mysterious," he writes. "Neither Steve Pinker [also an atheist] nor I can explain human subjective consciousness—what philosophers call *qualia*. In *How the Mind Works* Steve elegantly sets out the problem of subjective consciousness, and asks where it comes from and what's the explanation. Then he's honest enough to say, 'Beats the heck out of me.' That is an honest thing to say, and I echo it. We don't know. We don't understand it."[5]

Dawkins goes on to reassure his audience that while we do not understand consciousness in materialistic terms, neither did we understand life until a few decades ago. To support his point about life, he refers to the discovery of DNA. But this is like saying one understands the origin and nature of Mozart's Piano Concerto No. 21 after finally being allowed to lift the lid of a piano and see for the first time that there are hammers in there that strike a series of strings when the piano keys are struck. The person who sees for the first time the hammers and strings has learned something valuable about pianos and, by extension, about piano concertos, but unless he knows a great deal more about pianos and music theory and musical aesthetics and the creative process and the fact that piano concertos are composed by intelligent composers, he cannot be said to understand Mozart's piano concerto. In the same way, the discovery of the double-helical structure of DNA is an extraordinary achievement, but it's a far cry from understanding life. Fully understanding organic life involves grasping how it came to be, and that is precisely what we do not understand, at least not in materialistic terms—the hand-waving and grand claims of some scientific materialists notwithstanding.

The situation is really just the opposite of how Dawkins presents it. The scientific community thought it understood "simple life" and its origin. Now the origin-of-life community does not think so. Recall that, with all our advances in microscopy, chemistry, and molecular biology, leading synthetic chemist James Tour can say of the origin-of-life question, "I have asked all of my colleagues—National Academy members, Nobel Prize winners—I sit with them in offices. Nobody understands this. So if your professors say it's all worked out, if your teachers say it's all worked out, they don't know what they're talking about."[6]

It bears repeating: When Darwin first published *The Origin of Species*, the origin of the first one-celled life seemed like a simple matter. One-celled organisms were thought to be quite simple, and in any case, life appeared to spontaneously generate from non-life all the time. But the idea of spontaneous generation died a swift death at the hands of Louis Pasteur shortly after Darwin's *Origin* appeared. And the idea that one-celled life is simple has endured a slow, thorough dismembering in the generations that followed. So today geneticist Michael Denton can aptly describe even the smallest bacterial cell as "a veritable micro-miniaturized factory containing thousands of exquisitely designed pieces of intricate molecular machinery, made up altogether of one hundred thousand million atoms, far more complicated than any machinery built by man and absolutely without parallel in the non-living world."[7]

Denton speaks of a factory. I like to use my house as an example. Everything started when I was still living in Switzerland. It began with the thoughts of my architect friend. We instructed him to plan a typical Scandinavian wooden house. The drawings landed on my table and we offered some suggestions before accepting the plan. Then began the detailed planning: materials, electricity, ventilation, heating system, water systems, colors, ovens, windows, roof material, cupboards, fireplace, household machinery and of course, since it is Finland, a sauna. Then came my cousin, a general contractor, to organize and guide the building project. Everything started with a general plan and moved then to details. My house, like modern houses generally, has many layers of information and technology. The interior and exterior paint alone are a result of decades of development. The complexity of the cell is also layered in this way, only more so.

Factories and houses, both information-rich, are intelligently designed. Cells also are information-rich and appear designed. Indeed, cells are far more information-rich and information-dense than houses or human factories, and are designed with the capacity to do what no house or human factory can do—make copies of themselves that, in turn, make copies of themselves, etc. We bio-engineers—and indeed engineers generally—still have much to learn from the designs of life. There is now an entire discipline dedicated to this: biomimetics.

Add to this what was underscored in the previous chapter: Just fifteen years ago my pro-evolution colleagues at science conferences would express confidence that molecular biology had the genome mostly figured out, whereas today there is growing awareness that the genome and the cell are so sophisticated that we have barely scratched the surface. This growing understanding of life has pointed more and more insistently toward intelligent design. And I am convinced this trend will only continue. Organic life is information-rich, and its cause appears to be much the same as we find in our uniform experience with other information-rich artifacts. Its cause is an author, a coder, a composer, and one whose intelligence vastly outstrips ours.

I am convinced, as well, that this recognition will make biology more, not less, fruitful in the future. Consider again the Rosetta stone. The recognition that it was an intelligently designed object that needed decoding opened the doors to Egyptology on a totally different level. Similarly, a design perspective spurs fruitful lines of inquiry in the investigation of biological information, particularly when one regards the designer as not merely intelligent but ingenious. What is the purpose of the genetic art formerly considered junk under the neo-Darwinian paradigm? How much information is outside DNA? How are cell operations coordinated and regulated? How are organisms programmed to react to the environment? What are the limits of micro-variation? Can species borders be defined on a molecular level? What is the meaning of orphan genes? How much can we reprogram cells? One can approach these questions within a materialistic framework, but they make much more sense from a design perspective, and are especially well accommodated if one goes a step further and thinks, what ingenious techniques were used in the architecture of this or that living form that we perhaps do not yet grasp?

The Icon of Materialism

THE IDEA that scientific discoveries in biology have actually strengthened the design argument, and that the design perspective may prove especially fruitful, will seem counterintuitive to some. One reason for this is that our expectations about the progress of history have been shaped by a progress narrative that took root in the nineteenth century and is now conventional wisdom in some quarters, even in the face of mounting contrary evidence.

Biologist Jonathan Wells writes about the *icons of evolution*—popular evidences for evolution that get recycled in textbooks despite having been widely debunked even by mainstream evolutionists. Undergirding all of these icons is what we can call *the icon of materialism,* an icon that rescues Darwinian materialism any time it's in a jam.

Like the icons of evolution, the icon of materialism lives on even though it has been thoroughly debunked. It was given formal structure by the nineteenth-century French philosopher Auguste Comte. Comte asserted that investigation of the natural world has evolved through three stages.[8] In the theological phase, the mysterious doings of the gods are invoked to explain natural phenomena such as floods or plagues. In the metaphysical stage, natural phenomena may be explained by reference to abstract entities (such as the forms of Plato or the final causes of Aristotle). In the third and mature phase, natural phenomena are explained strictly by reference to natural laws or material processes.[9]

This view, presented either formally or informally, is one that certain atheists love to repeat. As a narrative, it runs like this: *Man used to chalk up every natural mystery to the gods—lightning bolts, illnesses, you name it. They stuffed a god into the gaps of their knowledge, shrugged, and moved on. The god of the gaps was a busy god. But as time went on, one scientific discovery after another filled in the gaps, shrinking the god of the gaps. The moral of the story: Even when the evidence seems to point to a designer, hold out for a purely materialistic, designer-free explanation. One is sure to come along—sooner or later.*

The tale is a grand one. It's also a myth. The myth says all the action has been in one direction—design explanations collapsing in the face of purely materialistic, designer-free explanations, and never the other way around. But that's untrue. Things have gone the other way around. For one, and as we saw above, it's gone the other way on the origin-of-life question. Years ago, scientists thought they had a good and tidy materialistic explanation for the origin of the first living cell. Today, we are light years from a materialistic explanation and the "micro-miniaturized factory" that is the single cell—bustling with robotics and a sophisticated information-processing system—looks for all the world like an intelligently designed system.

Thus, scientific discoveries have collapsed a trusted no-designer explanation for the origin of life, and bolstered the design explanation. The so-called "god of the gaps"... grew.

Another example, also noted earlier: In the nineteenth century, the smart money in science was on the view that we don't need to explain how the universe came to be because, well, it has always been. But discoveries in physics and astronomy put an end to this static-eternal model of the universe, and cosmologists now generally agree that our universe did have a beginning. So, what many thought never happened and didn't need explaining—the origin of the universe—suddenly cried out for an explanation. Then scientists began to uncover what today is widely known as *fine-tuning*: The laws and constants of physics and chemistry appear fine-tuned to allow for life. If the strength of gravity, or electromagnetism, or of the strong or weak nuclear force, or the speed of light—on and on the list goes—if any of these were even a tiny bit different, you couldn't get atoms beyond hydrogen and helium. You couldn't get life-essential carbon and water. You couldn't get stars and moons and planets. And without these, no life.

The fine-tuning is so striking that even committed atheists have abandoned ordinary appeals to chance. Instead they say there must be countless universes—a multiverse—and ours is just one of the lucky ones right for life. These other universes are, by definition, undetectable, so believing in the multiverse requires a faith commitment. Plus, the multiverse would itself need to be fine-tuned in order to spit out even the occasional universe capable of supporting life,[10] so the hypothesis merely moves the fine-tuning problem off-stage. It doesn't solve it.

Some physicists see the fine-tuning evidence pointing in a different direction. "Intelligent design, as one sees it from a scientific point of view, seems to be quite real," commented Nobel Laureate Charles Townes. "This is a very special universe: it's remarkable that it came out just this way."[11]

And this from another Nobel Laureate, astrophysicist Arno Penzias: "Astronomy leads us to a unique event, a universe which was created out of nothing, one with the very delicate balance needed to provide exactly the conditions required to permit life, and one which has an underlying (one might say 'supernatural') plan."[12]

So are these physicists "giving up on science," as some would claim? Not at all. Being open to the possibility of a designing intelligence isn't giving up on science or rationality or the experimental method. Rather, it's giving up on the myth of the ever-shrinking god of the gaps. It's letting the book of nature tell its own story, and following the story—the evidence—where it leads.

Nor is being open to the possibility of design a "science stopper," because it doesn't require us to reflexively assume design every time we do not understand some natural phenomenon. A phenomenon may be the direct product of design, or it may have arisen from natural processes alone, such as the famous crater in Arizona apparently formed by an asteroid meteorite strike. Such flexibility is in contrast to the dedicated materialist who must reflexively assume that blind material forces are the ultimate cause for everything in nature.

Voyage and Return

To be sure, scientific investigators keep discovering new ways that material forces cause and shape various things in nature. But atheists don't own the insight that we live in a world with underlying physical laws. Far from it. The idea was encouraged by the Christian belief that nature is the rational and orderly work of a divine mind, a cosmic lawgiver. That faith spurred Christians such as Copernicus, Galileo, and Kepler to go looking for the underlying laws. They looked for them, found them, and in the process launched the scientific revolution. It's no coincidence, after all, that the scientific revolution took place in Christian Europe, a revolution moreover with roots deep in the Middle Ages.[13]

Materialism's icon of the ever-shrinking God of the gaps obscures this historical reality, and layers myth upon myth. One of these is the myth that the Christian Middle Ages irrationally clung to belief in a flat Earth. This false history was born during the Enlightenment and given impetus by later histories, including Washington Irving's highly romanticized account of Christopher Columbus and a popular history by anti-religious propagandist John Draper, who also propagated the larger

FIGURE 12.2—Drawing from a university textbook from 1230, *De Sphaera Mundi*. It taught that Earth is a globe.

trope of a war between science and religion as part of a campaign to depict Christianity as backward and irrational.[14] Draper's work is titled *History of the Conflict Between Religion and Science*, and in it he asserts that in Columbus's day, the Papal Government's "traditions and policy forbade it to admit any other than the flat figure of the earth."[15] But this is pure fiction. In fact, virtually all educated people of the Middle Ages understood that the world was round.

Western thinkers had long known that the Earth is a globe and even made impressively accurate guesses at its rough circumference. They did so using principles of geometry and by measuring the distance at which a given ship's mast disappeared beneath the horizon. Those who opposed Columbus's attempted voyage around the world did so because they had much more accurately estimated the circumference of the globe, and quite reasonably feared that Columbus and all his sailors would die of thirst far out in the Atlantic Ocean.

Columbus was filled with confidence partly because he badly under-estimated the circumference of the earth. This led him to think that he could sail across the Atlantic and reach India well before running out food and water. The existence of the New World at roughly the spot where he thought he would reach India saved him and his ships from ruin. But neither Columbus nor any other educated person of his era thought the earth was flat.

It's true that Catholic Church authorities did resist the sun-centered model of the solar system for a time, but they did so as much from a com-mitment to Aristotelian/Ptolemaic science as from a particular reading of the Bible. And almost to a man the founders of the major branches of modern science were Christian—some Catholic and some Protestant. As Copernicus explained, he was seeking to uncover "the mechanism of the universe, wrought for us by a supremely good and orderly creator."[16] Kepler went even further. The laws of nature "are within the grasp of the human mind," he wrote, because "God wanted us to recognize them by creating us after his own image so that we could share in his own thoughts."[17]

The Theological Roots of Science

CHRISTIANS INVENTED modern science, but a later generation discard-ed science's fertile theological soil and insisted that science trade only in theories that fit materialism and atheism. They even redefined science as atheistic. Harvard geneticist Richard Lewontin frankly admits this. "We take the side of science *in spite* of the patent absurdity of some of its constructs... *in spite* of the tolerance of the scientific community for unsubstantiated just-so stories, because we have a prior commitment, a commitment to materialism," he writes (emphasis in original). And he continues:

> It is not that the methods and institutions of science somehow compel us to accept a material explanation of the phenomenal world, but, on the contrary, that we are forced by our a priori adherence to material causes to create an apparatus of investigation and a set of concepts that produce material explanations, no matter how coun-

ter-intuitive, no matter how mystifying to the uninitiated. Moreover, that materialism is absolute, for we cannot allow a Divine Foot in the door.[18]

The most basic of those unsubstantiated stories is the myth of an ever-shrinking god of the gaps. The myth ignores major developments in origin-of-life studies, physics, and astronomy. It ignores the reality that in significant areas, the evidence for intelligent design is not shrinking, but growing.

The renowned NASA astronomer and agnostic Robert Jastrow understood as much. He wrote that for the unbelieving scientist, confronted by the evidence of fine-tuning and a cosmic beginning, "the story ends like a bad dream. He has scaled the mountains of ignorance, he is about to conquer the highest peak; as he pulls himself over the final rock, he is greeted by a band of theologians who have been sitting there for centuries."[19]

A contemporary of Darwin's, the English poet Gerard Manley Hopkins, once wrote, "The world is charged with the grandeur of God. / It will flame out, like shining from shook foil."[20] Some are willing to go there when they register the evidence for design flashing forth from our collective study of life and the universe. Others are not. For those who remain unsure, I leave you with a modest invitation: Take at least that first step on the journey that I began so many decades ago as a young, slightly arrogant scientist committed to modern evolutionary theory. That first step is a modest one, a step through the door of a paradigm and onto an open path whose end point I was unsure of. The first step was the decision simply to follow the evidence wherever it led.

Will your adventure unfold as mine did? Will you be sure of your final destination? Not at all. But that, after all, is the nature of an adventure.

ENDNOTES

INTRODUCTION

1. David Berlinski, *The Devil's Delusion* (New York: Basic Books, 2008), 183–4.
2. "Smokers Assured in Industry Study," *The New York Times*, Aug. 17, 1964, accessed Oct. 6, 2017, http://www.nytimes.com/1964/08/17/smokers-assured-in-industry-study.html?_r=0.
3. "Pellagra in the United States of America," *History of Pellagra*, accessed Oct. 11, 2017, http://historyofpellagra.weebly.com/pellagra-in-the-us.html.
4. Jonathan Wells, *Zombie Science: More Icons of Evolution* (Seattle: Discovery Institute Press, 2017), 15–16.
5. Ian Dunham et al., "An Integrated Encyclopedia of DNA Elements in the Human Genome," *Nature* 489 (Sep. 2012): 57–74, doi:10.1038/nature11247.
6. Dan Graur et al., "On the Immortality of Television Sets: 'Function' in the Human Genome According to the Evolution-Free Gospel of ENCODE," *Genome Biology and Evolution* 5 no. 3 (March 2013): 578–590, doi:10.1093/gbe/evt028. https://www.ncbi.nlm.nih.gov/pubmed/23431001.
7. Richard Feynman, *Seeking New Laws* (Cambridge, Massachusetts: MIT Press, 1967), 156.

1. SUSPICIONS AWOKEN

1. *The Philebus of Plato*, trans. F. A. Paley (London: George Bell & Sons, 1873), 38.
2. Veikko Sorsa et al., *Lukion Biologia* (Helsinki, Finland: WSOY, 1974), 219.
3. Ernst Haeckel *Natürliche Schöpfungs-Geschichte* (Berlin: Reimer, 1868), 184. English version: *The History of Creation*, trans. E. Ray Lankester (New York: D. Appleton & Co., 1876).
4. Willy Ley, *Exotic Zoology* (New York: Viking Press, 1959), 409–11.
5. Charles Darwin to Joseph Hooker, February 1, 1871, in the University of Cambridge's Darwin Correspondence Project, https://www.darwinproject.ac.uk/letter/DCP-LETT-7471.xml.
6. André Brack, "Introduction," in *The Molecular Origins of Life: Assembling Pieces of the Puzzle*, ed. André Brack (Cambridge, England: Cambridge University Press, 1998).

7. Siegfried Scherer and Reinhard Junker, *Evolution - Ein Kritisches Lehrbuch* (Gießen, Germany: Weyel, 2013), 90-108.

8. George Gaylord Simpson, "The World into Which Darwin Led Us," *Science* 131, no. 3405 (1 April 1960): 966, http://science.sciencemag.org/content/131/3405/966.

9. Malcolm Dixon and Edwin C. Webb, *Enzymes*, 2nd edition (London: Longmans, 1964), 656–663.

10. Albert L. Lehninger, *Biochemistry* (New York: Worth Publishers, 1970), 769–792.

11. The Lehninger textbook described the experiments of Sidney Fox, which focused on processes considered important early steps in the origin of life. Fox had used pure amino acids and heated them in non-aqueous conditions. He obtained protein-like molecules that were not soluble in water but formed spherical structures suggested to be the precursors of cells.

12. The published research undermining Miller's assumptions about the early atmosphere is legion. A relatively recent and prominent paper of this sort is Dustin Trail, E. Bruce Watson, and Nicholas D. Tailby, "The Oxidation State of Hadean Magmas and Implications for Early Earth's Atmosphere," *Nature* 480 (Dec. 2011): 79–82, doi:10.1038/nature10655.

13. Stanley Miller's experiment was not unique at the time. A German scientist named Walther Loeb had conducted a similar experiment already in 1913, but his experiment was forgotten because it was not linked to the excitement surrounding origin-of-life research.

14. Stanley Miller and Leslie Orgel, *The Origin of Life on the Earth* (Englewood Cliffs, NJ: Prentice Hall, 1974), 144.

15. Robert Shapiro, *Origins: A Skeptic's Guide to the Creation of Life on Earth* (New York: Summit Books, 1986).

16. Robert Shapiro, "Small Molecule Interactions Were Central to the Origin of Life," *The Quarterly Review of Biology* 81, no. 2 (June 2006): 105–126, doi. org/10.1086/506024.

17. Stuart Kauffman, *At Home in the Universe: The Search for the Laws of Self-Organization and Complexity* (New York: Oxford University Press, 1995).

18. Marcel-Paul Schützenberger, "The Miracles of Darwinism," interview by *Origins & Design*, *Origins & Design* 17, no. 2 (Spring 1996), http://www.arn.org/docs/odesign/od172/schutz172.htm.

19. Klaus Dose and Anke Klein, "Response of *Bacillus subtilis* Spores to Dehydration and UV Irradiation at Extremely Low Temperatures," *Origins of life and evolution of the biosphere* 26, no. 1 (February 1996): 47–59.

20. Scherer and Junker, *Evolution - Ein Kritisches Lehrbuch*, 90-108.

21. Syozo Osawa, *Evolution of the Genetic Code* (New York: Oxford University Press, 1995), 177. On page 83 of the book, in the caption to a photo of an 11th-century plaque showing God and "the creation of the animals," Osawa adds, "The Creator and all these creatures use the common 'universal' genetic code." Since the Creator pictured in the plaque is highly anthropomorphized, we assume the comment is a tongue-in-cheek suggestion that the Creator is just one more member of the biological realm.

22. Fred Hoyle, *Facts and Dogmas in Cosmology and Elsewhere* (New York: Cambridge University Press, 1982), 12–13, quoted in Shapiro, *Origins*, 208.

23. James Tour, "The Origin of Life: An Inside Story—2016 Lectures," The Pascal Lectures on Christianity and the University, accessed Oct. 18, 2017, https://youtu.be/_zQXgJ-dXM4?t=3m6s (quotation begins at 3:06 of lecture).

24. Risto Nieminen, "Nano Ja Bio," *Helsingin Sanomat*, August 14, 2007.

25. A. E. Wilder-Smith, *Man's Origin, Man's Destiny* (Wheaton, Ill.; Harold Shaw, 1968), original German version: *Herkunft und Zukunft des Menschen* (Brunnen Verlag, 1966).

26. Paul S. Moorhead and Martin M. Kaplan, *Mathematical Challenges to the Neo-Darwinian Interpretation of Evolution*, The Wistar Institute monograph no. 5 (Wistar Institute Press, 1966). The Symposium proceedings are no longer available but parts of the presentations can be found here, accessed Nov. 8, 2017, http://www.evolutionnews.org/2006/07/mathematicians_and_evolution002387.html.

27. Marcel-Paul Schützenberger, "The Miracles of Darwinism," accessed Oct.6, 2017, http://www.arn.org/docs/odesign/od172/schutz172.htm.

28. For a survey and critique of attempts to model blind evolution via computer simulation, see Robert J. Marks II, William A. Dembski, and Winston Ewert, *Introduction to Evolutionary Informatics*, (Hackensack, New Jersey: World Scientific Publishing, 2017).

29. Douglas D. Axe, "Estimating the Prevalence of Protein Sequences Adopting Functional Enzyme Folds," *Journal of Molecular Biology* 341 (2004): 1295–1315.

30. Werner Arber, "The Existence of a Creator Represents a Satisfactory Solution," in *Cosmos, Bios, Theos: Scientists Reflect on Science, God, and the Origins of the Universe, Life, and Homo Sapiens*, ed. Henry Margenau and Roy Abraham Varghese (Peru, Ill.: Open Court Publishing, 1992), 141–143.

2. Fossilized Materialism

1. Siegfried Scherer and Reinhard Junker, *Evoluutio—Kriittinen Analyysi*, ed. Matti Leisola (Lahti: Datakirjat, 2000).

2. Thomas Lepeltier, *Vive le Créationisme! Point de Vue d'un Évolutionniste* (Editions de l'Aube, 2009), 59, quoted in Tapio Puolimatka, *Tiedekeskustelun Avoimuuskoe* (Helsinki: Uuusi Tie, 2010), 97.

3. Michael Ruse, "Is Darwinism a Religion?" *Huffington Post*, September 20, 2011, accessed Aug. 11, 2017, http://www.huffingtonpost.com/michael-ruse/is-darwinism-a-religion_b_904828.html.

4. Franklin Harold, *The Way of the Cell* (New York: Oxford University Press, 2001), 205.

5. Carl F. von Weizsäcker, *The Relevance of Science: Creation and Cosmogony* (New York: Harper and Row, 1964), 102.

6. Robert B. Laughlin, *A Different Universe* (New York: Basic Books, 2005), 168–169. Laughlin's use of the phrase "not even wrong" echoes theoretical physicist Wolfgang Pauli's use of the term, which has passed into common usage in science writing to

refer to a theory so speculative and muddled that one cannot rigorously examine or test it.

7. Scott C. Todd, "A View from Kansas on that Evolution Debate," correspondence to *Nature* 401 (September 30, 1999): 423, doi:10.1038/46661.

8. Jonathan Wells, *Zombie Science: More Icons of Evolution* (Seattle: Discovery Institute Press, 2017), 238. See also his earlier book on the subject, *Icons of Evolution* (Washington, D.C.: Regnery, 2001).

9. Stephen Meyer, *Darwin's Doubt: The Explosive Origin of Animal Life and the Case for Intelligent Design* (New York: HarperOne, 2013), 7, 20–22.

10. Stephen Meyer, *Darwin's Doubt*, 23–24. See also Louis Agassiz, "Evolution and the Permanence of Type," *Atlantic Monthly* 33 (1874): 92–101.

11. Robert L. Carroll, "Towards a New Evolutionary Synthesis," *Trends in Ecology & Evolution* 15 (2000): 27–32, doi:10.1038/npg.els.0001660.

12. Niles Eldredge and Stephen Jay Gould, "Punctuated Equilibria: An Alternative to Phyletic Gradualism," *Models in Paleobiology*, ed. Thomas J. M. Schopf (San Francisco: Freeman Cooper & Co., 1972), 250.

13. Stephen Meyer, *Darwin's Doubt*, 151. See James Valentine and Douglas Erwin, "Interpreting Great Developmental Experiments: The Fossil Record," *Development as an Evolutionary Process*, ed. by R. A. Raff and E. C. Raff (New York: Liss, 1987), 96.

14. Douglas Erwin and James Valentine, *The Cambrian Explosion: The Construction of Animal Biodiversity* (Greenwood Village, CO: Roberts and Company Publishers, 2013), 416.

15. Louis Agassiz, *Geological Sketches*: Vol. 1 (Boston: Ticknor and Fields, 1866), 22.

3. Students Begin Listening

1. Matti Leisola, *Evoluutio: Sattuman Uskonto* (self-published, 1977).

2. Julian Huxley, *Evolution in Action* (New York: Harper & Brothers, 1953), 272.

3. Maurice Caullery, *Genetics and Heredity*, translated by Mark Holloway (New York: Walker and Co., 1964), 10.

4. Robert P. Mortlock, D. D. Fossitt, W. A. Wood, "A Basis for Utilization of Unnatural Pentoses and Pentitols by *Aerobacter aerogenes*," 572-579, *Proceedings of the National Academy of Sciences* USA 54, no. 2 (1965): http://www.pnas.org/content/54/2/572.full.pdf+html.

4. Professors and Presidents React

1. Outi Rastas, "Kosmoksen Lottovoitto vai Kemiallinen Prosessi? Esko Valtaoja Pohtii Elämän Salaisuutta," *Kemia-Kemi* (2002, no. 9): 24. My response appeared in a subsequent issue: Matti Leisola, "Menninkäisiä Etsimässä," *Kemia-Kemi* (2003, no. 2), 43.

2. Matti Leisola, "Elämän Synnyn Arvoitus," *Ajankohtainen* 2 (1979): 4–7.

3. Kalervo V. Laurikainen, "Kehitysoppi on Tieteen Järjestelmä jolle Luomisteoria on vain Rasite," *Kristityn Vastuu* (Helsinki), March 1, 1980.

4. Esko Länsimies and Markku Myllykangas, "Keskustelua: Tiedeyliopistot on Puhdistettava Taikauskosta," *Acatiimi* (2010, no. 4), http://www.acatiimi. fi/9_2010/09_10_13.php. We responded in a letter published in the subsequent issue of the journal: Tapio Puolimatka and Matti Leisola, "Malttia Puhdistusintoon," *Acatiimi* (2011, no. 1), http://www.acatiimi.fi/1_2011/01_11_12. php.

5. Eero Bäckman, "Evoluutio—Kreationismi—Luomisusko; Keskustelu Suomalaisessa Lehdistössä 1981," C-sarja, no. 14 (Tampere: Kirkon Tutkimuslaitos, 1983).

6. Matti Leisola, "Evoluutioteorian Maailmankatsomuksellinen Luonne," *SKLS: n Vuosikirja* 13 (1991): 27–46.

7. Theodosius Dobzhansky, *Genetics and the Origin of Species* (New York: Columbia University Press, 1937), 12.

8. Del Ratzsch, *Nature, Design and Science: The Status of Design in Natural Science* (Albany, New York: SUNY Press, 2001), 98.

9. Emil Zuckerkandl and Linus Pauling, "Evolutionary Divergence and Convergence in Proteins," in *Evolving Genes and Proteins: A Symposium*, ed. Vernon Bryson and Henry J. Vogel (New York: Academic Press, 1965), 101.

10. Liliana Dávalos et. al, "Understanding Phylogenetic Incongruence: Lessons from Phyllostomid Bats," *Biological Reviews of the Cambridge Philosophical Society* 87 (2012): 991-1024, doi:10.1111/j.1469-185X.2012.00240.x. PMID:22891620.

11. Leonidas Salichos and Antonis Rokas, "Inferring Ancient Divergences Requires Genes with Strong Phylogenetic Signals," *Nature* 497 (May 16, 2013): 327–31, doi:10.1038/nature12130.

12. Emily Singer, "A New Approach to Building the Tree of Life," *Quanta*, June 4, 2013, accessed September 29, 2017, https://www.quantamagazine.org/a-new-approach-to-building-the-tree-of-life-20130604/.

13. James A. Shapiro, "Genome System Architecture and Natural Genetic Engineering in Evolution," *Annals of the New York Academy of Sciences* 870 (1999): 23–35, doi:10.1111/j.1749-6632.1999.tb08862.x.

14. Anne-Maria Mikkola et al., Äidinkieli ja Kirjallisuus (Helsinki, Finland: WSOY, 1998), 488.

15. Julian Huxley, *Evolution in Action* (New York: Harper & Bros, 1953), 36, 54–55.

16. Søren Løvtrup, "Macroevolution and Microevolution—Macromutations and Micromutations," *Rivista di Biologia* 80 (1987): 349–353.

17. For more on peppered moths and finch beaks as icons of evolution, see Jonathan Wells, *Zombie Science: More Icons of Evolution* (Seattle, Washington: Discovery Institute Press, 2017), 63–7, 67–71.

18. Nathan Ellis, Susan Ciocci, and James German, "Back Mutation Can Produce Phenotype Reversion in Bloom Syndrome Somatic Cells," 167–73, *Human Genetics*, 108, no. 2 (February 2001), doi.10.1007/s004390000447.

19. Michael Behe, *The Edge of Evolution: The Search for the Limits of Darwinism*, (New York: Free Press, 2007).

20. Michael Behe, "Experimental Evolution, Loss of-Function Mutations, and "the First Rule of Adaptive Evolution," *The Quarterly Review of Biology* 85, no. 4 (2010).

5. Publishers Hesitate

1. Michael Ruse and Edward O. Wilson, "The Evolution of Ethics," 50–52, *New Scientist*, October 17, 1985.

2. Matti Leisola and Niklas von Weymarn, *Bioprosessitekniikka* (Helsinki, Finland: WSOY, 2002), 416–417.

3. Siegfried Scherer and Reinhard Junker, *Evolution - Ein Kritisches Lehrbuch* (Gießen: Weyel, 2013).

4. Siegfried Scherer and Reinhard Junker, *Evoluutio – Kriittinen Analyysi*, ed. Matti Leisola (Lahti: Datakirjat, 2000).

5. Juuso Räsänen, *Kristityn Vastuu*, August 3, 2000.

6. Heli Karhumäki, "Kehittynyt vai Luotu," *Ilta-Sanomat*, August 19, 2000, A17.

7. Erkki A. Kauhanen, "Keskustelu Kertoo Teorian Voimasta," *Helsingin Sanomat*, August 26, 2000, https://www.hs.fi/tiede/art-2000003906624.html.

8. Quoted in "100 Scientists, National Poll Challenge Darwinism," Discovery Institute's Critique of PBS's Evolution, September 24, 2001, accessed Sept. 26, 2017, http://www.reviewevolution.com/press/pressRelease_100Scientists.php. (This citation was not in the original letter to the editor.)

9. A. E. Wilder-Smith, *Die Naturwissenschaften Kennen Keine Evolution - Experimentelle und Theoretische Einwände Gegen die Evolutionstheorie* (Basel: Schwabe & Co., 1980).

10. Anto Leikola, "Saarna Kehitysoppia Vastaan," *Helsingin Sanomat*, October 31, 1981.

11. A still more recent case of publisher intimidation concerned a 2011 scientific meeting on biological information held at Cornell University. Springer had agreed to publish the proceedings of the meeting and had even signed a contract, but backed out in 2012 after the Darwinist internet police attacked the book and its publication. For an account of the incident involving Springer, see Casey Luskin, "On the Origin of the Controversy over *Biological Information: New Perspectives*," *Evolution News & Science Today*, Aug. 19, 2013, accessed Nov. 1, 2017, http://www.evolutionnews. org/2013/08/on_the_origin_o_3075521.html.

12. Ilkka Niiniluoto, *Tiede, Filosofia ja Maailmankatsomus* (Helsinki: Otava, 1984), 358.

13. William Dembski, *Älykkään Suunnitelman Idea* (Lahti: Datakirjat, 2002).

14. George C. Williams, "A Package of Information," in *The Third Culture: Beyond the Scientific Revolution*, ed. John Brockman (New York: Simon & Schuster, 1995), 42–43.

15. Paul Davies, "How We Could Create Life," *The Guardian*, December 11, 2002, accessed Nov. 20, 2017, https://www.theguardian.com/education/2002/dec/11/highereducation.uk.

16. David Snoke, "Systems Biology as a Research Program for Intelligent Design," *BIO-Complexity* 3 (2014):1–11. doi:10.5048/BIO-C.2014.3.

17. Arthur Lander, "A Calculus of Purpose," *PLoS Biology* 2, no. 6 (2004): 0712, doi:10.1371/journal.pbio.0020164.

18. Pierre-Alain Braillard, "Systems Biology and the Mechanistic Framework," *History and Philosophy of the Life Sciences* 32 (2010): 43, http://www.ncbi.nlm.nih.gov/pubmed/20848805.

6. Broadcaster Bias

1. Dean H. Kenyon and Gary Steinman, *Biochemical Predestination* (New York: McGraw Hill, 1969).

2. A. E. Wilder-Smith, *The Creation of Life: A Cybernetic Approach to Evolution* (Chicago: Harold Shaw Publishers, 1970).

3. Fred Hoyle and Chandra Wickramasinghe, *Evolution from Space* (London: J. M. Dent & Sons, 1981).

4. Victor Meyer, "The Deep Waters of Evolution," YouTube video, October 5, 2011, accessed December 19, 2017, https://www.youtube.com/watch?v=GFQ3NczPazI.

5. Prisma Studio, "Darwinismi vs. Älykäs Suunnittelu," Yle Teema, February 2, 2005, accessed December 19, 2017, http://web.archive.org/web/20070510004146/http://www.yle.fi/teema/tiede/prisma_studio/id12369.html.

6. Prisma Studio, "Luomisoppi Vastaan Darwin," Yle Teema, October 1, 2007, accessed December 19, 2017, https://web.archive.org/web/20080319121400/http://ohjelmat.yle.fi/prisma/1_10_luomisoppi_vastaan_darwin.

7. In the interest of full disclosure, we note that the present work is published by Discovery Institute Press, and co-author Jonathan Witt is a Senior Fellow of Discovery Institute's Center for Science and Culture.

8. "PBS's Evolution Spikes Contrary Scientific Evidence, Promotes Its Own Brand of Religion," Discovery Institute's Critique of PBS's *Evolution*, accessed September 26, 2017, http://www.reviewevolution.com/viewersGuide/Evolution_00E.php.

9. "The Voyage That Shook the World," Con Dios Entertainment Pty Ltd and Fathom Media, Internet Archive, archived May 12, 2012, accessed December 19, 2017, https://web.archive.org/web/20120512103501/http://www.thevoyage.tv/.

10. Ted Baehr, "TV Review: The Voyage that Shook the World," Movie Guide, accessed September 22, 2017, https://www.movieguide.org/news-articles/tv-review-the-voyage-that-shook-the-world.html.

11. Seirian Sumner, Annual Question 2014: "What Scientific Idea is Ready for Retirement?" Response: "Life Evolves Via a Shared Genetic Toolkit," accessed October 20, 2017, https://www.edge.org/response-detail/25533.

12. John F. McDonald, "The Molecular Basis of Adaptation: A Critical Review of Relevant Ideas and Observations," *Annual Review of Ecological Systems* 14 (1983): 77–102.

13. "More Light is Cast on Epigenetics and Design," *Evolution News & Science Today*, June 10, 2013, accessed October 23, 2017, http://www.evolutionnews.org/2013/06/more_light_is_c073041.html.

14. Stephen C. Meyer, *Darwin's Doubt: The Explosive Origin of Animal Life and the Case for Intelligent Design* (San Francisco: HarperOne, 2013), 285. For more on this subject, see also Jonathan Wells, "Membrane Patterns Carry Ontogenetic Information that is Specified Independently of DNA," *BIO-Complexity* 2014, no.2, http://bio-complexity.org/ojs/index.php/main/article/view/BIO-C.2014.2.

7. The Church Evolves

1. John Shelby Spong, *The Bishop's Voice: Selected Essays*, ed. Christine M. Spong (New York: Crossroad Publishing, 1999), 226.

2. Eero Huovinen, *Suomen Evankelis-Luterilaisen Kirkon Kristinoppi – Katekismus* (Helsinki: Edita, 2000), 36.

3. Philip Johnson, *Defeating Darwinism by Opening Minds* (Downers Grove, Illinois: InterVarsity Press, 1997), 18–19.

4. John West, "Is Darwinian Evolution Compatible with Religion?" Discovery Institute, May 1, 2009, accessed September 30, 2017, https://www.discovery.org/a/9721.

5. Ibid.

6. "New Poll Reveals Evolution's Corrosive Impact on Beliefs about Human Uniqueness," Discovery Institute, April 5, 2016, accessed September 30, 2017, https://www.discovery.org/scripts/viewDB/filesDB-download. php?command=download&id=12031.

7. Paul Feyerabend, *Farewell to Reason* (New York: Verso, 1996), 260, 264.

8. Mikko Juva, letter to the editor, *Helsingin Sanomat*, June 6, 1981.

9. Mikko Juva, "Uskonnonvastaisen Naturalismin Tunkeutumisesta Suomen Sivistyselämään," *Suomalaisen Kirkkohistoriallisen Seuran Vuosikirja*: 1949–1950 (Helsinki: Finnish Society of Church History, 1951).

10. Mikko Heikka, "Moraalikatoon ei ole enää Varaa," *Suomen Kuvalehti*, April 10, 2009, Internet Archive, archived January 15, 2011, https://web.archive.org/ web/20110115051454/https://suomenkuvalehti.fi/blogit/eri-mielta/mikko-heikka-moraalikatoon-ei-ole-enaa-varaa.

11. Mikko Heikka, "Paneeko Tiede Jumalan Viralta?" *Suomen Kuvalehti*, September 5, 2008, Internet Archive, archived October 17, 2011, accessed December 19, 2017, https://web.archive.org/web/20111017015455/http://suomenkuvalehti.fi/blogit/ eri-mielta/heikka-paneeko-tiede-jumalan-viralta.

12. E. Yaroslavsky, *Landmarks in the Life of Stalin* (London: Lawrence & Wishart, 1942), 9. Simon Sebag Montefiore, in *Young Stalin* (New York: Alfred A. Knopf, 2007), 49, also relates this chapter in Stalin's life, and the additional details in Montefiore's telling confirm the central role that Darwin's theory of evolution played in drawing Stalin toward atheism:

When he [Stalin] was about thirteen, Lado Ketskhoveli took him to a little bookshop where he paid a five kopeck subscription and borrowed a book that was probably Darwin's *Origin of Species*. Stalin read it all night, forgetting to sleep, until Keke [Stalin's mother] found him.

"Time to go to bed," she said. "Go to sleep—dawn is breaking."

"I loved the book so much, Mummy, I couldn't stop reading…" [ellipses in original] As his reading intensified, his piety wavered.

One day Soso [Stalin] and some friends, including Grisha Glurjidze, lay on the grass in town talking about the injustice of there being rich and poor when he amazed all of them by suddenly saying: "God's not unjust, he doesn't actually exist. We've been deceived. If God existed he'd have made the world more just."

"Soso! How can you say such things?' exclaimed Grisha.

"I'll lend you a book and you'll see.'" He presented Grisha with a copy of Darwin."

13. "Interview with William Provine," in *Expelled: No Intelligence Allowed*, directed by Logan Craft (2008; Premise Media Corporation).

14. Eero Junkkaala, *Alussa Jumala Loi… Luomisusko ja Tieteellinen Maailmankuva* (Kauniainen: Perussanoma Oy, 2013).

15. "Peer-Reviewed Articles Supporting Intelligent Design," Discovery Institute: Center for Science and Culture, accessed December 19, 2017, https://www.discovery.org/id/peer-review/.

16. Frank Tipler, "Refereed Journals: Do They Insure Quality or Enforce Orthodoxy?" in *Uncommon Dissent: Intellectuals Who Find Darwinism Unconvincing*, ed. William Dembski (Wilmington, Delaware: ISI Books, 2004), 115–30.

17. Rafael D'Andrea and James P. O'Dwyer, "Can Editors Save Peer Review from Peer Reviewers?" *PLOS One* 12, no. 10 (October 9, 2017): e0186111, http://journals.plos.org/plosone/article?id=10.1371/journal.pone.0186111.

18. Matti Leisola et al., "Aromatic Ring Cleavage of Veratryl Alcohol by *Phanerochaete chrysosporium*," *FEBS Letters* 189 (1985): 267–270, doi:10.1016/0014-5793(85)81037-1.

19. Toshiaki Umezawa and Takayoshi Higuchi, "Mechanism of Aromatic Ring Cleavage of ⊠-O-4 Lignin Substructure Models by Lignin Peroxidase," *FEBS Letters* 218, no. 2 (June 29, 1987): 255–60 (see Ref. 4), http://www.sciencedirect.com/science/article/pii/0014579387810578. Also, see the in-depth study my group did later: Stephan D. Haemmerli, Hans E. Schoemaker, Harald W. H. Schmidt, and Matti S. A. Leisola, "Oxidation of Veratryl Alcohol by the Lignin Peroxidase of *Phanerochaete chrysosporium* Involvement of Activated Oxygen," *FEBS Letters*, Vol. 220, no. 1 (August 10, 1987): 149–54, http://onlinelibrary.wiley.com/doi/10.1016/0014-5793(87)80893-1/full.

20. Stephan Haemmerli, Matti Leisola, and Armin Fiechter, "Polymerization of Lignins by Ligninases from *Phanerochaete chrysosporium*," *FEMS Microbiology Letters* 35, no. 1 (1986): 33–36, https://doi.org/10.1111/j.1574-6968.1986.tb01494.x.

21. Jacob Troller et al., "Crystallization of the Lignin Peroxidase from the White-Rot Fungus *Phanerochaete chrysosporium*," *Nature Biotechnology* 6 (1988): 571–573, doi:10.1038/nbt0588-571.

22. Mary Schweitzer et al., "Soft Tissue Vessels and Cellular Preservation in Tyrannosaurus Rex," *Science* 307 (2005): 1952–1955, doi:10.1126/science.1108397.

23. Barry Yeoman, "Schweitzer's Dangerous Discovery," *Discover*, April 27, 2006, accessed November 7, 2017, http://discovermagazine.com/2006/apr/dinosaur-dna.

24. Warren B. Hamilton, "Archean Tectonics and Magmatism," *International Geology Review* 40 (1998): 1–39.

25. R. L. Armstrong, "The Persistent Myth of Crustal Growth," *Australian Journal of Earth Sciences* 38 (1991): 613–630.

26. Günther Blobel, quoted in Lawrence K. Altman, "Rockefeller U. Biologist Wins Nobel Prize for Protein Cell Research," *New York Times*, October 12, 1999, accessed November 20, 2017, http://www.nytimes.com/1999/10/12/nyregion/rockefeller-u-biologist-wins-nobel-prize-for-protein-cell-research.html.

27. The *PLOS ONE* Staff, "Retraction: Biomechanical Characteristics of Hand Coordination in Grasping Activities of Daily Living," *PLOS ONE* 11, no. 3 (March 4, 2016): e0151685, https://doi.org/10.1371/journal.pone.0151685.

8. "Rationalists" Behaving Irrationally

1. "Skepsis ry on myöntänyt 2008 Huuhaa-palkinnon Kustannus Oy Uusi tielle," Skepsis Ry, accessed November 11, 2017, http://www.skepsis.fi/HuuhaaPalkinnot.

2. "Skepsis ry on myöntänyt 1989 Huuhaa-palkinnon Werner Söderström Osakeyhtiölle," Skepsis Ry, accessed November 11, 2017, http://www.skepsis.fi/HuuhaaPalkinnot.

3. Matti Leisola, "Pyhää Lehmää Potkittiin," *Kemia-Kemi* 2 (2005): 39, http://www.kemia-lehti.fi/kemia-kemi-22005/.

4. Esko Valtaoja, "Kosmoksen Siruja," *Ursan Julkaisuja 122*, (Helsinki, Finland: Tähtitieteellinen yhdistys Ursa, 2010) 212.

5. "A Scientific Dissent from Darwinism," Discovery Institute, accessed December 15, 2017, http://www.discovery.org/scripts/viewDB/filesDB-download.php?command=download&id=660.

6. There is a bright side to this particular story about my time at Cultor. The recommendation not to hire me due to my work in Christian student ministry was made to the director, who happened to be very positive about Christianity. When I joined the company my immediate boss asked me "to keep my mouth shut" on topics related to worldview or I would never be successful in the company. Ironically, he was later fired and I was put in his place. During my nine years with the company, the coffee-table discussions were often about evolution and its mechanism. At the farewell party I got a congratulations card with the words, "From research director to professor—such a jump is not possible without intelligent design!"—which shows that my views were known to everyone in the company.

7. P. G. Humber, "Debating Dawkins," 1–4, *Creation Matters* 8 (2003).

8. Ibid.

9. Nathaniel Comfort, "Genetics: Dawkins, Redux," *Nature* 525 (September 10, 2015): 184–185, doi:10.1038/525184a.

10. Casey Luskin, "Richard Dawkins on Darwin-Doubting Undergraduate Student: 'Little Fool' Is a 'Pathetic Little Idiot,'" June 17, 2012, *Evolution News & Science Today*, accessed November 22, 2017, https://evolutionnews.org/2012/06/richard_dawkins_3/.

11. Richard Dawkins, review of *Blueprints: Solving the Mystery of Evolution*, by Richard Milton, *New Statesman*, August 28, 1992.

12. U. S. House of Representatives Committee on Government Reform, "Intolerance and the Politicization of Science at the Smithsonian," Staff Report Prepared for The Hon. Mark Souder, Chairman, Subcommittee on Criminal Justice, Drug Policy and Human Resources (December 11, 2006), http://www.discovery.org/f/1489. See also U. S. Office of Special Counsel, Letter to Richard Sternberg, August 5, 2005, http://www.discovery.org/f/1488.

13. Richard Sternberg, "Smithsonian Controversy," RichardSternberg.org, accessed November 7, 2017, http://www.richardsternberg.com/smithsonian.php.

14. "Bibliographic and Annotated List of Peer-Reviewed Publications Supporting Intelligent Design," Center for Science and Culture, Discovery Institute, July 2017, accessed November 7, 2017, http://www.discovery.org/scripts/viewDB/filesDB-download.php?command=download&id=10141.

15. Richard Dawkins, *The Selfish Gene* (Oxford: Oxford University Press, 1976), 47.

16. Leslie E. Orgel and Francis Crick, "Selfish DNA: The Ultimate Parasite," *Nature* 284 (1980): 604–607, doi:10.1038/284604a0.

17. Douglas J. Futuyma, *Evolution* (Sunderland, MA: Sinauer Associates, 2005), 48–49.

18. Michael Shermer, *Why Darwin Matters: The Case Against Intelligent Design* (New York: Henry Holt and Company, 2006), 74–75.

19. Jerry Coyne, *Why Evolution is True* (New York: Viking, 2009), 66–67.

20. John C. Avise, *Inside the Human Genome: A Case for Non-Intelligent Design* (Oxford: Oxford University Press, 2010), 82, 115.

21. Jonathan Wells, *The Myth of Junk DNA* (Seattle: Discovery Institute Press, 2011).

22. Richard Sternberg, "On the Roles of Repetitive DNA Elements in the Context of a Unified Genomic-Epigenetic System," *Annals of the New York Academy of Sciences* 981 (December 2002): 154–88.

23. Ann Gauger, Ola Hössjer, and Colin Reeves, "Evidence for Human Uniqueness," 475-502, *Theistic Evolution: A Scientific, Philosophical, and Theological Critique,* ed. J. P. Moreland et. al, (Wheaton, Illinois: Crossway, 2017), 475.

24. James A. Shapiro and Richard von Sternberg, "Why Repetitive DNA is Essential to Genome Function," 1–24, *Biological Reviews* 80 (2005), doi:10.1017/S1464793104006657; Richard von Sternberg and James A. Shapiro, "How Repeated Retroelements Format Genome Function," 108–116, *Cytogenetic and Genome Research* 110 (2005), doi:10.1159/000084942.

25. James A Shapiro, "Genome Organization and Reorganization in Evolution," *Annals of the New York Academy of Sciences* 981 (December 2002): 111–134, doi:10.1111/j.1749-6632.2002.tb04915.x.

9. Colleagues Dare to Explore

1. Matti Leisola, "Bioscience, Bioinnovations, and Bioethics," in *Green Gene Technology: Research in Areas of Social Conflict*, ed. Armin Fiechter and Christof Sautter (Berlin: Springer, 2007), 41–56. The volume is one of the *Advances in Biochemical Engineering/Biotechnology* series, ed. T. Scheper.

2. Michael Denton, *Evolution: A Theory in Crisis* (Chevy Chase, MD: Adler & Adler, 1986).

3. D. R. Mills, R. L. Peterson, and Sol Spiegelman, "An Extracellular Darwinian Experiment with Self-Duplicating Nucleic Acid Molecule," *Proceedings of the National Academy of Sciences USA* 58, no. 1 (1967): 217–224, http://www.pnas.org/content/58/1/217.full.pdf+html.

4. Branko Kozulić, "Proteins and Genes, Singletons and Species," *ViXra.org*, accessed November 8, 2017, http://vixra.org/pdf/1105.0025v1.pdf.

5. Branko Kozulić and Matti Leisola, "Have Scientists Already Been Able to Surpass the Capabilities of Evolution?" *ViXra.org*, April 17, 2015, accessed December 19, 2017, http://vixra.org/pdf/1504.0130v1.pdf.

6. I met Professor Mortlock in the first meeting of *The International Society of Rare Sugars* (http://isrs.kagawa-u.ac.jp/society.html) in Japan. We both belonged to the international committee for this organization. Another member of the committee was origin-of-life researcher Arthur Weber, a colleague of the late Stanley Miller, famous for his origin-of-life experiments in the 1950s. Weber is a senior research scientist at the SETI Institute (http://www.seti.org/users/arthur-weber), which searches for radio-signal evidence of intelligent life in star-systems beyond ours. In a conversation I had with him at the meeting, Weber said he considered Miller's experiments as meaningless. He thought that a plausible route to living cells was via sugars but admitted that we are far from a solution. I invited him to join the editorial board of *BIO-Complexity* to discuss these scientific problems openly, but like many other evolutionists, he responded with silence.

7. Robert P. Mortlock, D. D. Fossitt, and W. A. Wood, "A Basis for Utilization of Unnatural Pentoses and Pentitols by *Aerobacter aerogenes*," *Proceedings of the National Academy of Sciences USA* 54, no 2 (1965): 572–579, http://www.pnas.org/content/54/2/572.full.pdf+html.

8. Bob Holmes, "Bacteria Make Major Evolutionary Shift in the Lab," *New Scientist*, June 9, 2008, accessed November 7, 2017, http://www.newscientist.com/article/dn14094-bacteria-make-major-evolutionary-shift-in-the-lab.html#.Ublu9NhjHyY.

9. Michael Behe, *The Edge of Evolution: The Search for the Limits of Darwinism* (New York: Free Press, 2007), 142.

10. Sebastien Wielgoss et al., "Mutation Rate Dynamics in a Bacterial Population Reflect Tension between Adaptation and Genetic Load," *Proceedings of the National Academy of Sciences USA* 110 (2013): 222–227, doi:10.1073/pnas.1219574110.

11. Michael Behe, "Richard Lenski and Citrate Hype—Now Deflated," *Evolution News & Science Today*, May 12, 2016, accessed Nov. 1, 2017, https://evolutionnews.org/2016/05/richard_lenski/. See also D. J. Van Hofwegen, C. J. Hovde, and S. A. Minnich, "Rapid Evolution of Citrate Utilization by *Escherichia coli* by Direct Selection Requires *citT* and *dctA*," *Journal of Bacteriology* 198, no. 7 (April 2016): 1022–34, doi:10.1128/JB.00831-15.

12. Casey Luskin, "Hype from *New Scientist* Aside, Lenski's *E. coli* Research Shows Evolution of Nothing New," *Evolution News & Science Today*, accessed November 4, 2017, https://evolutionnews.org/2015/06/hype_from_new_s/.

13. Alan E. Linton, review of *The Triumph of Evolution and the Failure of Creationism*, by Niles Eldredge, *Times Higher Education Supplement* (April 20, 2001), 29.

14. Barry G. Hall, "The EBG System of *E. coli:* Origin and Evolution of a Novel Beta-Galactosidase for the Metabolism of Lactose," *Genetica* 118 (2003): 143–156, http://www.ncbi.nlm.nih.gov/pubmed/12868605.

15. Joakim Näsvall et al., "Real-Time Evolution of New Genes by Innovation, Amplification, and Divergence," *Science* 338, no. 6105 (2012): 384–387, doi:10.1126/science.1226521.

16. University of California-Davis, "Evolution of New Genes Captured," *ScienceDaily*, October 22, 2012, accessed December 28, 2017, www.sciencedaily.com/releases/2012/10/121022145340.htm.

17. Ian Chant, "Better, Faster, Stronger: Evolution of New Genes Seen in Lab for First Time," *The Mary Sue*, October 22nd, 2012, accessed January 3, 2018, https://www.themarysue.com/evolution-new-genes/.

18. Ann K. Gauger et al. "Reductive Evolution Can Prevent Populations from Taking Simple Adaptive Paths to High Fitness," *BIO-Complexity* 2010, no. 2 (April 23, 2010): 1–9, doi:10.5048/BIO-C.2010.2.

19. Krishnendu Chatterjee et al. "The Time Scale of Evolutionary Innovation," *PLoS Computational Biology* 10, no. 9 (2014): e1003818, doi:10.1371/journal.pcbi.1003818.

10. Mechanisms Malfunction

1. Juha Apajalahti and Matti Leisola, Yeast Strains for the Production of Xylitol, US Patent 6271007, 1994.

2. "Biochemical Pathway Maps," ExPasy Bioinformatics Resource Portal, accessed December 19, 2017, http://web.expasy.org/pathways/.

3. Anu Harkki et al., Recombinant Method and Host for Manufacture of Xylitol, US Patent 5631150, 1995.

4. The Seventh International Symposium on the Life Sciences, held in Japan on November 14–17, 1988.

5. Matti Leisola and Ossi Turunen, "Protein Engineering: Opportunities and Challenges," *Applied Microbiology and Biotechnology* 75, no. 6 (2007): 1225–1232, doi:10.1007/s00253-007-0964-2.

6. Koen Beerens, "Characterization and Engineering of Epimerases for the Production of Rare Sugars," PhD diss., Ghent University, 2013, https://biblio.ugent.be/record/3125525.

7. "Reconstruction of Prehistoric DNA Refutes Criticism on Theory of Evolution," December 12, 2012, Internet Archive, archived February 21, 2013, accessed December 19, 2017, https://web.archive.org/web/20130221014153/http://www.ugent.be/en/news/bulletin/prehistoric-dna.htm.

8. Karin Voordeckers et al., "Reconstruction of Ancestral Metabolic Enzymes Reveals Molecular Mechanisms Underlying Evolutionary Innovation through Gene Duplication," *PLOS Biology* (December 11, 2012), http://www.plosbiology.org/article/info%3Adoi%2F10.1371%2Fjournal.pbio.1001446.

9. Richard T. Halvorson, "Confessions of a Skeptic," *Harvard Crimson*, April 7, 2003, accessed November 11, 2017, http://www.thecrimson.com/article/2003/4/7/confessions-of-a-skeptic-does-our/.

10. Douglas Axe, "Belgian Waffle," *Evolution News & Science Today,* January 18, 2013, accessed November 11, 2017, http://www.evolutionnews.org/2013/01/belgian_waffle068421.html.

11. "Sieni Sulki Öljyhanat," *Helsingin Yliopisto Ajankohtaista,* July 5, 2012, Internet Archive, archived December 2, 2012, https://web.archive.org/web/20121202211300/http://www.helsinki.fi/ajankohtaista/uutisarkisto/7-2012/5-12-10-11.html.

12. Dimitrios Floudas et al. "The Paleozoic Origin of Enzymatic Lignin Decomposition Reconstructed from 31 Fungal Genomes," *Science* 336 (2012): 1715–1719. *doi*:10.1126/science.1221748.

13. Matti Leisola, Ossi Pastinen, and Douglas D. Axe, "Lignin—Designed Randomness," *BIO-Complexity* 2012, no. 3 (2012):1–11. doi:10.5048/BIO-C.2012.3.

14. Douglas D. Axe, "The Case Against a Darwinian Origin of Protein Folds," *BIO-Complexity* 2010, no. 1 (2010): 1–12. doi:10.5048/BIO-C.2010.1.

15. Evelyn Fox Keller, *The Century of the Gene* (Cambridge, Massachusetts: Harvard University Press, 2000), 130–131.

16. Ann Gauger et al. "Reductive Evolution Can Prevent Populations from Taking Simple Adaptive Paths to High Fitness," *BIO-Complexity* 2010, no. 2: 1–9. doi:10.5048/BIO-C.201.

17. Leisola et al., "Lignin—Designed Randomness."

18. Matti Leisola, "Lignin Is and Remains Enigmatic," *Evolution News & Science Today,* July 27, 2012, accessed November 28, 2012, http://www.evolutionnews.org/2012/07/lignin_is_and_r062611.html.

19. Luciano Brocchieri and Samuel Karlin, "Protein Length in Eukaryotic and Prokaryotic Proteomes," *Nucleic Acids Research* 33 (2005): 3390–3400. doi:10.1093/nar/gki615.

20. Sean V. Taylor et al., "Searching Sequence Space for Protein Catalysts," *Proceedings of the National Academy of Sciences* 98, no. 19 (2001): 10596–10601. doi:10.1073/pnas.191159298.

21. Hubert P. Yockey, "A Calculation of the Probability of Spontaneous Biogenesis by Information Theory," *Journal of Theoretical Biology* 67 (1977): 377–398, available at http://www.sciencedirect.com/science/article/pii/0022519377900443?via%3Dihub.

22. John F. Reidhaar-Olson and Robert T. Sauer, "Functionally Acceptable Substitutions in Two ☒-helical Regions of ☒ Repressor," *Proteins* 7, no. 4 (1990): 306–316. doi:10.1002/prot.34007040.

23. Douglas D. Axe, "Estimating the Prevalence of Protein Sequences Adopting Functional Enzyme Folds," *Journal of Molecular Biology* 341 (2004): 1295–1315. doi:10.1016/j.jmb.2004.06.058.

24. Kirk Durston and David K. Chiu, (2012) "Functional Sequence Complexity in Biopolymers," in *The First Gene: The Birth of Programming, Messaging and Formal Control,* ed. David L. Abel (New York: LongView Press, 2012), 117–134.

25. Leisola and Turunen, "Protein Engineering: Opportunities and Challenges."

26. Hairong Xiong et al., "Engineering the Thermostability of *Trichoderma reesei* endo-1,4-ⵣ-xylanase II by Combination of Disulphide Bridges," *Extremophiles* 8, no. 5 (2004): 393–400. doi:10.1007/s00792-004-0400-9.

27. In our example the code for cysteine is tgc. However, a code tgt also results in cysteine. If we allow this change as well, the probability doubles.

28. Michael Behe, *The Edge of Evolution: The Search for the Limits of Darwinism* (New York: Free Press, 2007), 146.

29. Olga Khersonsky, Cintia Roodveldt and Dan S. Tawfik, "Enzyme Promiscuity: Evolutionary and Mechanistic Aspects," *Current Opinion of Chemical Biology* 10 (2006): 498–508. doi:10.1016/j.cbpa.2006.08.011.

30. Relevant here is my work on the well-known industrial enzyme glucose isomerase, used in fructose manufacturing. My research team and I found that the enzyme has many minor side activities, something we discovered by using large enzyme amounts in the reaction. These weak side activities were carried out by the same catalytic site in the enzyme. Various sugar related to glucose, fructose, and xylose could fit—although less efficiently—in the same active pocket in the enzyme. See Ossi Pastinen, Kalevi Visuri, Hans E. Schoemaker and Matti Leisola, "Novel Reactions of Xylose Isomerase from *Streptomyces riginosusub*," *Enzyme and Microbial Technology* 25, no. 8–9 (November 1999): 695–700, https://doi.org/10.1016/S0141-0229(99)00100-3. Also see Antti Vuolanto, Ossi Pastinen, Hans E. Schoemaker and Matti Leisola, "C-2 Epimer Formation of Tetrose, Pentose and Hexose Sugars by Xylose Isomeras," *Biocatalysis and Biotransformation* 20, no. 4 (2002): 235–240, https://doi.org/10.1080/10242420290029463. We then improved one of the side activities by protein engineering 3-fold: Johanna Karimäki et al., "Engineering the Substrate Specificity of Xylose Isomerase," *Protein Engineering, Design and Selection* 17, no. 12 (December 2004): 861–869, https://doi.org/10.1093/protein/gzh099.

31. Douglas Axe and Ann Gauger, "Model and Laboratory Demonstrations that Evolutionary Optimization Works Well Only If Preceded by Invention—Selection Itself Is Not Inventive," *BIO-Complexity* 2015, no. 2 (2015): 1–13. doi:10.5048/BIO-C.2015.2.

32. Francisco J. Blanco, Isabelle Angrand and Luis Serrano, "Exploring the Conformational Properties of the Sequence Space between Two Proteins wih Different Folds: An Experimental Study," *Journal of Molecular Biology* 285 (1999): 741–753, http://www.ncbi.nlm.nih.gov/pubmed/9878441.

33. Ann K. Gauger and Douglas D. Axe, "The Evolutionary Accessibility of New Enzyme Functions: A Case Study from the Biotin Pathway," *BIO-Complexity* 2011, no. 1 (2011): 1–17. doi:10.5048/BIO-C.2011.1.

34. Jesse D. Bloom et al., "Protein Stability Promotes Evolvability," *Proceedings of the National Academy of the Sciences* 103 (2006): 5869–5874, doi:10.1073_pnas.0510098103.

35. Dan Tawfik quoted in Rajendrani Mukhopadhyay, "Close to a Miracle:' Researchers are Debating the Origins of Proteins," *American Society for Biochemistry and Molecular Biology* 12, no. 9 (2013): 13, accessed November 17, 2017, http://www.asbmb.org/asbmbtoday/asbmbtoday_article.aspx?id=48961.

11. THE CHASM WIDENS

1. Manfred Mortell et al., "Physician 'Defiance' Towards Hand Hygiene Compliance: Is There a Theory-Practice-Ethics Gap?" *Journal of the Saudi Heart Association*, 25, no. 3 (July 2013): 203–208, doi: http://dx.doi.org/10.1016/j.jsha.2013.04.003.

2. Antoine Lavoisier, quoted in Douglas McKie, *Antoine Lavoisier: The Father of Modern Chemistry* (Philadelphia: J. P. Lippincott Company, 1936), 230.

3. James B. Conant, *Science and Common Sense* (New Haven, CT: Yale University Press, 1962).

4. Fred Hoyle and Chandra Wickramasinghe, *Evolution from Space* (London: J. M. Dent & Sons, 1981), 135.

5. Richard Goldschmidt, *The Material Basis of Evolution* (New Haven, CT: Yale University Press, 1940), 438.

6. Stephen J. Gould, "The Return of Hopeful Monsters," *Natural History* 86 (1979): 22–30.

7. Casey Luskin, "Darwinian Evolution Gets Left Behind," *Evolution News & Science Today*, November 1, 2012, accessed November 13, 2017, http://www.evolutionnews.org/2012/11/darwinian_evolu065911.html.

8. Graham Budd, quoted in John Whitfield, "Biological Theory: Postmodern Evolution," *Nature* 455 (2008): 281–284, doi:10.1038/455281a.

9. Antony Flew, "My Pilgrimage from Atheism to Theism," interview by Gary Habermas, *Philosophia Christi* 6, no. 2 (2004): 197–211.

10. Stephen Meyer, *Signature in the Cell: DNA and the Evidence for Intelligent Design* (New York: HarperCollins, 2009).

11. Thomas Nagel, *Mind and Cosmos: Why the Materialist Neo-Darwinian Conception of Nature Is Almost Certainly False* (Oxford: Oxford University Press, 2012), 10.

12. Tapio Puolimatka, "Evoluutioteoriaa on Opetettava Kriittisesti Avoimella Tavalla," *Helsingin Sanomat*, November 15, 2008, Internet Archive, archived September 15, 2010, https://web.archive.org/web/20100915080506/http://www.hs.fi/paakirjoitus/artikkeli/Evoluutioteoriaa+on+opetettava+kriittisesti+avoimella+tavalla/1135241111292.

13. Hanna Kokko and Katja Bargum, *Kutistuva Turska* (Helsinki: WSOY, 2008).

14. For more on the theological, philosophical, and aesthetic problems with the bad-design arguments in origins science, see Cornelius Hunter, *Darwin's God: Evolution and the Problem of Evil* (Grand Rapids, MI: Brazos Press, 2001) and Benjamin Wiker and Jonathan Witt, *A Meaningful World: How the Arts and Sciences Reveal the Genius of Nature* (Downers Grove, IL: IVP Academic, 2006).

15. Stephen J. Gould, *Evolution as Fact and Theory in Hen's Teeth and Horse's Toes* (New York: W. W. Norton & Company, 1980), 254–255.

16. Richard Dawkins, *The God Delusion* (New York: Bantam Books, 2006).

17. Kari Enqvist, *Kuoleman ja Unohtamisen Aikakirjat* (Helsinki: WSOY, 2009), 126–127.

18. Stephen Meyer, *Darwin's Doubt* (New York: HarperOne, 2013), x.

19. Philip Ball, "DNA: Celebrate the Unknowns," *Nature* 496 (2013): 419–420, doi:10.1038/496419a.

20. Jeffrey Norris, "Brain Development is Guided by Junk DNA That Isn't Really Junk," University of California San Francisco News Center, April 15, 2013, accessed November 14, 2017, http://www.ucsf.edu/news/2013/04/105126/brain-development-guided-junk-dna-isn%E2%80%99t-really-junk.

21. Vicent Pelechano, Wu Wei, and Lars M. Steinmetz, "Extensive Transcriptional Heterogeneity Revealed by Isoform Profiling," *Nature* 497 (May 2, 2013): 127–131, doi:10.1038/nature12121.

22. Michael Ruse, "Is Darwinism a Religion?" *Huffington Post*, September 20, 2011, accessed Aug. 11, 2017, http://www.huffingtonpost.com/michael-ruse/is-darwinism-a-religion_b_904828.html.

23. Stuart Kauffman is the best-known member of the Santa Fe group. He makes an extended case for the self-organization model in *At Home in the Universe: The Search for the Laws of Self-Organization and Complexity* (New York: Oxford University Press, 1995).

24. For a lengthier critique of self-organization, see Chapter 15 of Stephen Meyer's *Darwin's Doubt: The Explosive Origin of Animal Life and the Case for Intelligent Design* (New York: HarperCollins, 2014).

25. Jeffrey H. Schwartz, "Homeobox Genes, Fossils, and the Origin of Species," *Anatomical Record* 257 (1999): 15–31, http://www.pitt.edu/~jhs/articles/homeobox_genes.pdf.

26. Wallace Arthur, "Internal Factors in Evolution: The Morphogenetic Tree, Developmental Bias, and Some Thoughts on the Conceptual Structure of Evo-Devo," 343–63, in *Conceptual Change in Biology: Scientific and Philosophical Perspectives on Evolution and Development,* ed. Alan C. Love, Boston Studies in the Philosophy and History of Science 307 (Springer Verlag, 2014), 350. Arthur goes on to explore possible ways around this problem, but the attempts are tentative and laced with additional concessions. For a critical look at Arthur's attempt to salvage evo-devo from the contrary research evidence, see Paul Nelson, "Accepting, or Rejecting, Common Descent Has Real Consequences for Biological Understanding," *Evolution News & Science Today*, accessed December 4, 2017, https://evolutionnews.org/2016/01/accepting_or_re/.

27. Eva Jablonka and Gal Raz, "Transgenerational Epigenetic Inheritance: Prevalence, Mechanisms, and Implications for the Study of Heredity and Evolution," *Quarterly Review of Biology* 84, no. 2 (2009): 131–76.

28. Parental non-coding RNA can, for example, influence the DNA copy number in protozoa. See Mariusz Nowacki et al., "RNA-Mediated Epigenetic Regulation of DNA Copy Number," *PNAS* 2010 107 (51) 22140-22144, doi:10.1073/pnas.1012236107. The total RNA injected into a fertilized egg can induce a heritable white tail phenotype in mice. See Minoo Rassoulzadegan et al., "RNA-Mediated Non-Mendelian Inheritance of an Epigenetic Change in the Mouse," *Nature* 441 (2006): 469–474, http://dx.doi.org/10.1038/nature04674.

29. James A. Shapiro, *Evolution: A View from the 21st Century* (Upper Saddle River, NJ: FT Press Science, 2011).

30. What follows is a by no means comprehensive list of evolutionists/sources challenging and/or revising the modern evolutionary synthesis: Eric H. Davidson, "Evolutionary Bioscience as Regulatory Systems Biology," *Developmental Biology* 357 (2011): 35–40, doi:10.1007/978-1-4614-3567-9_1; Douglas Erwin, "Macroevolution Is More than Repeated Rounds of Microevolution," *Evolution and Development* 2 (2000): 78–84, doi:10.1046/j.1525-142x.2000.00045.x; Gerry Webster and Brian Goodwin, *Form and Transformation: Generative and Relational Principles in Biology* (Cambridge: Cambridge University Press, 2011); Eugene V. Koonin, "*The Origin* at 150: Is a New Evolutionary Synthesis in Sight?" *Trends in Genetics* 25 (2009): 473–475, doi:10.1016/j.tig.2009.09.007; Lynn Margulis and Dorion Sagan, *Acquiring Genomes: A Theory of the Origin of Species* (New York: Basic Books, 2002); Armin P. Moczek, "On the Origins of Novelty in Development and Evolution," *BioEssays* 30 (2008): 432–447; Jan Sapp, "The Structure of Microbial Evolutionary Theory," *Studies in History and Philosophy of Biological and Biomedical Sciences* 38 (2007): 780–95, doi:10.1016/j.shpsc.2007.09.011; Neil H. Shubin and Charles R. Marshall, "Fossils, Genes, and the Origin of Novelty," *Paleobiology* 26 (2000): 324–340, doi:10.1093/molbev/msn282; Arlin Stoltzfus, "Mutationism and the Dual Causation of Evolutionary Change," *Evolution and Development* 8 (2006): 304–317, doi:10.1111/j.1525-142X.2006.00101.x; Günther Theissen, "Saltational Evolution: Hopeful Monsters are Here to Stay," *Theory in Biosciences* 128 (2009): 43–51, doi:10.1007/s12064-009-0058-z; James W. Valentine, *On the Origin of Phyla* (Chicago: University of Chicago Press, 2004); Andreas Wagner, "The Molecular Origins of Evolutionary Innovations," *Trends in Genetics* 27 (2011): 397–410, doi:10.1016/j.tig.2011.06.002; and Günter P. Wagner and Hans C. E. Larsson, "What Is the Promise of Developmental Evolution?" *Journal of Experimental Zoology, Part B: Molecular and Developmental Evolution* 300 (2003): 1–4.

12. Through a Doorway to Adventure

1. The theological and philosophical questions swirling around theistic evolution are quite involved. They are not the focus of this book. For an accessible point and counterpoint discussion, see J. B. Stump, ed., *Four Views on Creation, Evolution, and Intelligent Design* (Grand Rapids: Zondervan, 2017). For a multifaceted critique of theistic evolution, we recommend either of two anthologies, the first of moderate length and the second weighing in at more than a thousand pages: Jay W. Richards, ed., *God and Evolution: Protestants, Catholics, and Jews Explore Darwin's Challenge to Faith* (Seattle, Washington: Discovery Institute Press, 2010); and J. P. Moreland et al., eds., *Theistic Evolution: A Scientific, Philosophical, and Theological Critique* (Wheaton, Illinois: Crossway, 2017).

2. Graham Bell, *Selection: The Mechanism of Evolution* (New York: Chapman & Hall, 1997), 553.

3. Thomas Nagel, *Mind and Cosmos: Why the Materialist Neo-Darwinian Conception of Nature is Almost Certainly False* (New York: Oxford University Press, 2012), 53.

4. Jerry A. Fodor, "The Big Idea: Can There Be a Science of the Mind," *Times Literary Supplement*, July 3, 1992, 5.

5. Richard Dawkins and Stephen Pinker, "Is Science Killing the Soul," Guardian-Dillons Debate at the Westminster Central Hall (London), February 10, 1999,

in *Edge* 53 (April, 8, 1999), accessed December 7, 2017, https://www.edge.org/documents/archive/edge53.html.

6. James Tour, "The Origin of Life: An Inside Story," The 2016 Pascal Lectures on Christianity and the University, accessed Oct. 18, 2017, https://youtu.be/_zQXgJ-dXM4?t=3m6s (quotation begins at 3:06 of lecture).

7. Michael Denton, *Evolution: A Theory in Crisis* (Chevy Chase, Maryland: Adler & Adler, 1986), 250.

8. Auguste Comte, "Plan of the Scientific Operations Necessary for Reorganizing Society (Third Essay, 1822)," in Gertrud Lenzer, ed., *August Comte and Positivism: The Essential Writings* (New York: Transaction Publishers 1998), 9–70.

9. This section of the chapter, on the myth of the God of the shrinking gaps, and Comte's three stages, is adapted from Jonathan Witt, "The Icon of Materialism: Why Scientism's Cherished Progress Narrative Fails," March/April 2015, *Touchstone*, accessed December 5, 2015, http://www.touchstonemag.com/archives/article.php?id=28-02-040-f, and "The 'God of the Gaps' is Growing," July 14, 2017, *The Imaginative Conservative*, accessed Dec. 5, 2017, http://www.theimaginativeconservative.org/2017/07/god-gaps-growing-jonathan-witt.html.

10. As Robin Collins explains, focusing on what is probably the leading multiverse theory (though still highly speculative) "Even if an inflationary-superstring multiverse generator exists, it must have just the right combination of laws and fields for the production of life-permitting universes: if one of the components were missing or different, such as Einstein's equation or the Pauli Exclusion Principle, it is unlikely that any life-permitting universes could be produced. Consequently, at most, this highly speculative scenario would explain the fine-tuning of the constants of physics, but at the cost of postulating additional fine-tuning of the laws of nature." See "The Teleological Argument: An Exploration of the Fine-Tuning of the Universe," in William Lane Craig and J. P. Moreland, eds., *The Blackwell Companion to Natural Theology* (West Sussex, England: Blackwell, 2009), 265.

11. Charles Townes, interviewed by Bonnie Azab Powell, June 17, 2005, *UCBerkeley News*, accessed December 8, 2017, http://www.berkeley.edu/news/media/releases/2005/06/17_townes.shtml.

12. Arno Penzias, "Creation is Supported by All the Data So Far," in H. Margenau and R. A. Varghese, eds., *Cosmos, Bios, and Theos* (La Salle, Illinois: Open Court Press, 1992), 83.

13. For a debunking of the "Dark Ages" myth and an overview of the evidence that Medieval learning and innovation laid much of the groundwork for the scientific revolution (now conventional wisdom among historians of the Middle Ages), see Rodney Stark, *How the West Won: The Neglected Story of the Triumph of Modernity* (Wilmington, Delaware: ISI Books, 2015).

14. Jeffrey B. Russell, *Inventing the Flat Earth: Columbus and Modern Historians* (Westport, CT: Praeger, 1997).

15. John William Draper, *History of the Conflict between Religion and Science* (New York: D. Appleton and Company, 1875), 83.

16. Nicholaus Copernicus, *On the Revolutions of the Heavenly Spheres*, Preface and Book 1, translated by John F. Dobson and Selig Brodetsky, in *Occasional Notes of the Royal Astronomical Society*, No. 10, v. 2, May 1947), 149–173; Milton K. Munitz, ed.,

Theories of the Universe: From Babylonian Myth to Modern Science (New York: Simon & Schuster, 1965).

17. Kepler to Johann Georg Herwart von Hohenburg, in Carola Baumgardt, *Johannes Kepler: Life and Letters* (New York: Philosophical Library, 1951), 50.

18. Richard Lewontin, "Billions and Billions of Demons," January 9, 1997, *The New York Review of Books*, accessed December 8, 2017, http://www.nybooks.com/articles/1997/01/09/billions-and-billions-of-demons/.

19. Robert Jastrow, *God and the Astronomers*, 2nd edition (New York: W. W. Norton & Company, 1992), 107.

20. Gerard Manley Hopkins, "God's Grandeur," Poetry Foundation, https://www.poetryfoundation.org/poems/44395/gods-grandeur.

Image Credits

Chapter 1

1.1— Ossi Turunen, used with permission.

1.2—Ernst Haeckel, public domain.

1.3—Kimmo Pälikkö.

1.4—Ray Braun, used with permission, redrawn by Kimmo Pällikö.

1.5—William Ely Hill, public domain, via Wikimedia Commons. Originally appeared in *Puck*, v. 78, no. 2018 (1915 Nov. 6), 11.

1.6—Matti Leisola.

Chapter 2

2.1—Ray Braun, used with permission, redrawn by Kimmo Pällikö.

2.2—By Unidentified photographer [No restrictions], via Wikimedia Commons.

Chapter 3

3.1—Matti Leisola.

3.2—Kimmo Pällikö.

3.3—Kimmo Pällikö.

3.4—Kimmo Pällikö.

Chapter 4

4.1—Matti Leisola.

4.2—Siegfried Scherer, used with permission.

4.3—Janica Candolin, used with permission.

4.4—Courtesy of Studiengemeinschaft Wort und Wissen e. V.

4.5—Courtesy of Studiengemeinschaft Wort und Wissen e. V.

4.6—Courtesy of Studiengemeinschaft Wort und Wissen e. V.

Chapter 5

5.1—Matti Leisola.

Chapter 6

6.1—Kimmo Pällikö.

Chapter 7

7.1—By Eiyaha11 (Own work) [CC BY-SA 3.0 (https://creativecommons.org/licenses/by-sa/3.0)], via Wikimedia Commons.

7.2—Hannu Savonen, Uusi Tie Ltd., used with permission.

Chapter 8

8.1—Sammeli Juntunen, used with permission.

8.2—Laszlo Benzce.

Chapter 9

9.1—Matti Leisola.

9.2—Brigitte Schmidt, used with permission.

9.3—Braniko Kozulić, used with permission.

9.4—Kimmo Pällikö.

9.5—Kimmo Pällikö.

9.6—Kimmo Pällikö.

Chapter 10

10.1—Matti Leisola.

10.2—Koen Beerens, used with permission.

10.3—Ossi Turunen, used with permission.

10.4—Kimmo Pällikö.

10.5—Kimmo Pällikö.

10.6—Matti Leisola.

10.7—Matti Leisola.

10.8—Kimmo Pällikö.

10.9—Matti Leisola.

Chapter 11

11.1—Courtesy of Aamulehti.

Chapter 12

12.1—By Albeins [GFDL (http://www.gnu.org/copyleft/fdl.html) or CC BY-SA 4.0-3.0-2.5-2.0-1.0 (https://creativecommons.org/licenses/by-sa/4.0-3.0-2.5-2.0-1.0)], via Wikimedia Commons.

12.2—Public domain via Wikimedia Commons.

INDEX

Made in the USA
San Bernardino, CA
20 March 2018